AF074460

INDUSTRIAL PROBLEM SOLVING SIMPLIFIED

AN 8-STEP PROGRAM

Ralph R. Pawlak

apress®

Industrial Problem Solving Simplified: An 8-Step Program

Copyright © 2014 by Ralph R. Pawlak

This work is subject to copyright. All rights are reserved by the Publisher, whether the whole or part of the material is concerned, specifically the rights of translation, reprinting, reuse of illustrations, recitation, broadcasting, reproduction on microfilms or in any other physical way, and transmission or information storage and retrieval, electronic adaptation, computer software, or by similar or dissimilar methodology now known or hereafter developed. Exempted from this legal reservation are brief excerpts in connection with reviews or scholarly analysis or material supplied specifically for the purpose of being entered and executed on a computer system, for exclusive use by the purchaser of the work. Duplication of this publication or parts thereof is permitted only under the provisions of the Copyright Law of the Publisher's location, in its current version, and permission for use must always be obtained from Springer. Permissions for use may be obtained through RightsLink at the Copyright Clearance Center. Violations are liable to prosecution under the respective Copyright Law.

ISBN-13 (pbk): 978-1-4302-6577-1

ISBN-13 (electronic): 978-1-4302-6578-8

Trademarked names, logos, and images may appear in this book. Rather than use a trademark symbol with every occurrence of a trademarked name, logo, or image we use the names, logos, and images only in an editorial fashion and to the benefit of the trademark owner, with no intention of infringement of the trademark.

The use in this publication of trade names, trademarks, service marks, and similar terms, even if they are not identified as such, is not to be taken as an expression of opinion as to whether or not they are subject to proprietary rights.

While the advice and information in this book are believed to be true and accurate at the date of publication, neither the authors nor the editors nor the publisher can accept any legal responsibility for any errors or omissions that may be made. The publisher makes no warranty, express or implied, with respect to the material contained herein.

>President and Publisher: Paul Manning
>Acquisitions Editor: Jeff Olson
>Editorial Board: Steve Anglin, Mark Beckner, Ewan Buckingham, Gary Cornell, Louise Corrigan, Jonathan Gennick, Jonathan Hassell, Robert Hutchinson, Michelle Lowman, James Markham, Matthew Moodie, Jeff Olson, Jeffrey Pepper, Douglas Pundick, Ben Renow-Clarke, Dominic Shakeshaft, Gwenan Spearing, Matt Wade, Tom Welsh
>Coordinating Editor: Rita Fernando
>Copy Editor: Kezia Endsley
>Compositor: SPi Global
>Indexer: SPi Global
>Cover Designer: Anna Ishchenko

Distributed to the book trade worldwide by Springer Science+Business Media New York, 233 Spring Street, 6th Floor, New York, NY 10013. Phone 1-800-SPRINGER, fax (201) 348-4505, e-mail orders-ny@springer-sbm.com, or visit www.springeronline.com. Apress Media, LLC is a California LLC and the sole member (owner) is Springer Science + Business Media Finance Inc (SSBM Finance Inc). SSBM Finance Inc is a Delaware corporation.

For information on translations, please e-mail rights@apress.com, or visit www.apress.com.

Apress and friends of ED books may be purchased in bulk for academic, corporate, or promotional use. eBook versions and licenses are also available for most titles. For more information, reference our Special Bulk Sales–eBook Licensing web page at www.apress.com/bulk-sales.

Any source code or other supplementary materials referenced by the author in this text is available to readers at www.apress.com. For detailed information about how to locate your book's source code, go to www.apress.com/source-code/.

Apress Business: The Unbiased Source of Business Information

Apress business books provide essential information and practical advice, each written for practitioners by recognized experts. Busy managers and professionals in all areas of the business world—and at all levels of technical sophistication—look to our books for the actionable ideas and tools they need to solve problems, update and enhance their professional skills, make their work lives easier, and capitalize on opportunity.

Whatever the topic on the business spectrum—entrepreneurship, finance, sales, marketing, management, regulation, information technology, among others—Apress has been praised for providing the objective information and unbiased advice you need to excel in your daily work life. Our authors have no axes to grind; they understand they have one job only—to deliver up-to-date, accurate information simply, concisely, and with deep insight that addresses the real needs of our readers.

It is increasingly hard to find information—whether in the news media, on the Internet, and now all too often in books—that is even-handed and has your best interests at heart. We therefore hope that you enjoy this book, which has been carefully crafted to meet our standards of quality and unbiased coverage.

We are always interested in your feedback or ideas for new titles. Perhaps you'd even like to write a book yourself. Whatever the case, reach out to us at editorial@apress.com and an editor will respond swiftly. Incidentally, at the back of this book, you will find a list of useful related titles. Please visit us at www.apress.com to sign up for newsletters and discounts on future purchases.

The Apress Business Team

To my wife, Rose, who has always been my main support and companion. And also to our two sons, Glen and Ralph, who have exceeded all of our lofty expectations.

Contents

Preface ix
About the Author xi
Acknowledgments xiii
Introduction xv

Chapter 1: Define the Problem 1
Chapter 2: Define Fault Characteristics 9
Chapter 3: Construct a Concept Sheet 17
Chapter 4: Develop a Plan of Attack 35
Chapter 5: Collect Relevant Data 63
Chapter 6: Generate Clues 73
Chapter 7: Choose and Use Analysis Tools 89
Chapter 8: Use Innovative Analysis Tools 137
Chapter 9: Establish Consistent Work, Many-Level Reviews,
 and Certification 155
Chapter 10: Summary 167
Appendix A: Fractional Explained 169
Appendix B: Interaction Explained 177
Appendix C: Cracked-or-Broken Example 183
Appendix D: Torque-to-Turn Example 189
Appendix E: Sum of Extremes Test 197
Appendix F: Definitions 203

Index 207

Preface

Some readers might ask why they should invest the time needed to read this book. My hope is that the answer is provided in the following pages.

Foreign competition and job outsourcing have caused dislocations in many industries. Since economic uncertainty abounds across the world, most manufacturing facilities attempt to maximize their competitiveness. Due to the attempts to control costs, there are plant closings, employment stagnations, dislocations, and reductions. None of these actions is beneficial to the long-term viability of a manufacturing facility unless they are coupled with innovative changes that increase competitive capability. Consequently, managers evaluate each employee's attributes in efforts to retain talented workers who are believed to sustain the company's future.

Since affected individuals must yield to these changing conditions, education and experience on the job are primary attributes that can be used in the decision-making process. This may be true even in union-protected bargaining units if the opportunity presents itself. Therefore, it benefits all individuals to be viewed as contributing and irreplaceable employees.

There are many owners, managers, and technical and mechanical personnel who have excellent training in their respective fields, yet are without competence as trained problem-solvers. This book provides a simple method to be used independently or in conjunction with standard, sophisticated problem-solving tools. The methods and examples explained in this book stand alone and can be easily applied, without in-depth statistical competence or understanding.

When you use these methods, you are more likely to increase profits, decrease scrap, and improve the manufacturing process. Consequently, people who enact profitable improvements are more likely viewed as valuable assets and not as expendable.

After earning two degrees, one in industrial technology and the other in industrial engineering, I was assigned to act as a problem-solver in a high-volume automotive foundry. Upon receiving the position, the manager of quality called me into his office and asked me to improve the plant scrap quality level. He indicated that I had complete control of how my skills were used and that he expected to see reductions in waste and the other negative trends the plant was experiencing. And he wanted them *now!*

Preface

I was overjoyed at the opportunity, since I had a lot of academic theory from a variety of subjects. So I defined problems, recorded data, and created charts to indicate the current scrap levels by defect. Through this process, I found that charts alone do not solve problems.

Generally, I found that the major differences were assignable to different production lines with different causes for each line. On a daily basis, I found incorrect chemistries, mixed parts, cold iron, burn-in, core cuts, sand holes, shrinks, blows, shift differences, pattern differences, part differences, line downtime, and the foundry nemesis—porosity. There were so many variables that it was difficult to pick out the most significant ones. Consequently, I decided to rank those defects based on their effect on the plant's bottom line. I placed the cost data rankings in a chart to allow the most significant targets to be picked for reduction. I could not think of a better alternative.

There were some obvious defects that were easily corrected, but the majority of the defects could not be solved with the aid of the technical information that I had available. Book knowledge is a wonderful asset, but I lacked the real-world knowledge and the knack to determine the causes or conditions necessary to facilitate change. Over time, I gained the experience and developed that knack, and this book contains both, in distilled form for your benefit.

During that time, the most important lesson that I learned was that there must be agreement about the problem being discussed. It is not unusual to have eight or so people discussing a defect without even two of them agreeing on the specific description or nature of the problem.

For years, I had a gnawing desire to identify more specifically the tools needed to investigate and correct problems. Using the simple tools available didn't typically result in expedient solutions.

Working with various managers over the years, I realized that even they have difficulty recognizing and overcoming basics associated with problem-solving. Thus, this treatise is an attempt to provide the basic knowledge and skill.

I invite you to become a proficient problem-solver using the plan, examples, and worksheets provided in this book. With these tools, your efforts will be focused naturally on the actual cause of the problem. Follow the eight-step roadmap provided in the following pages and see how easy your journey to manufacturing problem resolution can become.

About the Author

Ralph R. Pawlak attended Erie Community College after being honorably discharged from the USMCR. Upon graduating with a degree in industrial technology, he attended the General Motors Institute in Flint, Michigan, where he earned a degree in industrial engineering. Although he has a master of education degree from the University of Buffalo, his heart lies in his engineering background. During a 25-year career with General Motors, he served as troubleshooter, engineer, supervisor, quality manager, and superintendent of plant engineering at various facilities in Buffalo, New York; Tonawanda, New York; Danville, Illinois; and Romulus, Michigan. He has also held positions in quality assurance at Fisher-Price, Vibratech Inc., and Continental Automotive.

Acknowledgments

My thanks to the following Apress/Springer Science + Business Media representatives who took my rudimentary thoughts and polished them into acceptable literary form.

- Jeff Olson, the executive editor, who facilitated the smooth transition.
- Rita Fernando, the coordinating editor, who made the transition painless.
- Kezia Endsley, who copyedited the manuscript superbly.
- SPi International, which helped improve the drawings and figures in the book.

Without their help and assistance, the information would not have been nearly as clear or acceptable.

Introduction

Most of the time, chronic manufacturing problems are ignored if their effect is tolerable. Only after the effect of a manufacturing problem spikes are the problems recognized and addressed. It would be a manager's Nirvana to have a magic wand to cause these aggravating problems to disappear. Unfortunately, the "wizardry" needed to systematically identify, study, and solve problems has not been achieved by the people most responsible for managing and solving industrial problems.

Process improvement is an ongoing activity that should be applied whenever your company learns a new lesson. The examples and illustrations in this book originate from the experience I gained working with more than 200 individual suppliers while in high-volume manufacturing facilities. When any discrepant condition affecting the process was identified, I included it in the body of the correction process with an assigned corrective action. This was done to prevent recurrences. These items were then systematically added to the Design Failure Modes and Effect Analysis (DFMEA) or Process Failure Modes and Effect Analysis (PFMEA),[1] and in the control plans[2] for current and future products.

These techniques have been applied effectively in the United States, Canada, Italy, Hong Kong, and China, and in facilities that manufactured toys, chemicals, electronics, castings, strollers, adolescent clothing, gears, engines, and other machined or assembled components. The processes used to detect and solve problems can be used anywhere to expedite industrial problem-solving resolutions.

Results of "common knowledge" methods depend upon the skills and training of the individual involved. Many individuals have not been trained in effective problem-solving techniques. A formal education allows you to develop cognitive skills, but does little to establish an efficient roadmap for guide those thrown into the problem-solving battle.

---

[1] The *DFMEA* defines the design considerations that will be applied to a product, whereas a *PFMEA* relates to the process you use to ensure you obtain the desired engineering and aesthetic results.

[2] A *control plan* is a specified method for organizing and managing a manufacturing or service process.

Introduction

I wrote this book to provide a simple map that can help anyone transform difficult technical problems into easily solved ones. The method involves applying creative steps to improve understanding while using the tools available. These tools enable those unfamiliar with a manufacturing process to attack problems, even when they don't fully understand the fundamentals of that process.

Some organizations have never attempted to improve efficiency, and therefore profits, by using a system to identify problems. As a result, unidentified problems may and probably have caused a significant reduction in bottom-line results. Therefore, applying corrective actions to eliminate unaddressed underlying quality problems will boost profits.

A Time-Tested Problem-Solving Technique

These six actions are necessary to scrutinize and resolve recognized problems:

1. Define the specific problem and condition (the characteristics or physical appearance of the subject) thoroughly. Evaluate problems by the dollar loss that they generate.

2. Verify that the process is operating as intended and specified. Conduct a review of the control plan and the operating procedures to ensure that the process is being managed as desired.

3. Observe the operation for detrimental or ruinous conditions. Determine if there is anything specific causing distress in the process.

4. Develop a check system for recognized manufacturing control items. Decide what is important to control so that the process is acceptable.

5. Conduct checks with more than one layer of inspection. Two or more people should conduct the checks individually.

6. React to the conditions that must be corrected or improved. Add corrective actions to the audit list if any item or condition is found that has or will have a negative effect on the process or the product.

Each of the six steps is discussed throughout the book. If these actions are used in conjunction with established, consistent work instructions, they will be instrumental in preventing errors, losses (scrap, rework, lost production, and so on), and decreases in customer satisfaction. However, this is only the beginning of the correction process. Customer expectations must also be fulfilled in order to increase future demand and profits. Consequently, any expectation not fulfilled may affect customer satisfaction and may adversely affect future profitability.

These steps are applicable to small as well as difficult problems. The methods I use to illustrate them include actual manufacturing examples and conditions, which provides relevance and clarity for understanding. When you use these practices, you can expect improved operations and profit. This is true whether the manager has professional problem-solving experience or not.

Roadmap to the Book

Although some of these tools are used in industry, the quality of their results depends on the individual using them. The following pages illustrate the power inherent in successful problem-solving techniques and provide the information you need to use them effectively.

Examples provide insight into the nature and use of steps available to achieve problem-solving effectiveness. They will help the novice or expert problem-solver alike.

Note You do not have to understand statistical methods and statistical analysis to apply the tools described in this book. The steps and tools can be easily applied to technical or service problems. In addition, you will find that they become easier to use with each application.

Some of the verbiage used in this book may not be familiar to you. These include acronyms and terms like FMEA, consistent work, job instructions, variable gauges, sporadic incidence, plan of attack, calibration, foreign material (FM), interaction, and concept diagram. Seeing these terms in context may be sufficient for understanding, but if they're still unclear, refer to Appendix F, "Definitions."

When you're faced with a problem, you normally compare five good and five bad samples to generate clues. But due to the need for simplicity, photograph clarity, and the necessary size of image display, I have reduced the sample size to three good and three bad in most cases. This reduced sample size still provides clarity for understanding. (See Appendix E, "Sum of Extremes Test," for the rationale used for this comparison.)

The contents follow a plan that provides you with the steps necessary to evaluate any problem. These actions begin with:

Chapter 1: Define the Problem (Step One). This chapter introduces you to the concept that you can evaluate and explain any problem if it's understood thoroughly. If you observe and define part characteristics adequately, it is easier to surmise the conditions that may have caused the characteristic. Further, the chapter introduces methods for collecting data and applying available tools to evaluate current problems.

Chapter 2: Define Fault Characteristics (Step Two). This chapter explains the methods for understanding the problem defect or condition by determining the characteristics that may be causing problems. You'll read about considerations that are useful in identifying problems, including defect scene characteristics and contrasts of two, both of which aid future analysis.

Chapter 3: Construct a Concept Sheet (Step Three). This chapter introduces the use of concept sheets, which list any conditions suspected to cause the detrimental condition. Case studies in welding, leaking, and broken component problems illustrate how to use concept sheets. You'll also become more proficient in using them over time.

Chapter 4: Develop a Plan of Attack (Step Four). This chapter identifies a method for moving toward a solution. It introduces the concept of interactions, which can adversely affect processes. Sketches and photographs illustrate problems that I have experienced and resolved. This chapters also explains how to collect and compare samples whenever a problem arises.

Chapter 5: Collect Relevant Data (Step Five). This chapter describes the types of data that you can collect. It differentiates between causative and quantifiable data and explains how to use both in rectifying problems. It explains how to use a visual rating system when numerical data can't be collected easily or at all. It also provides examples of constructing visual systems so that you can understand the difference between good and bad part criteria.

Chapter 6: Generate Clues (an Interlude). Although not one of the formal eight steps, this chapter contains other problem-solving tools useful for generating clues. It summarizes numerous tools I have used successfully with past problems. The chapter also describes tests with examples you can use to generate clues. These are the sum of extremes, duos, data ranking, and good versus bad comparisons, all of which you can use to generate clues or to verify that corrective actions were satisfactory.

Chapter 7: Choose and Use Analysis Tools (Step Six). This chapter introduces innovative tools to aid you in resolving problems. It proposes generating visual clues about how a component passes through a system with more than one flow path. It also introduces the use of other tools, such as the noise matrix layout and the cracked-and-broken worksheet.

Chapter 8: Use Innovative Analysis Tools (Step Seven). You may not be familiar with some of the tools that are presented in this chapter. You can augment the use of consistent work and review lists by creating visual comparison systems to attack your problem. These include flow charts, matrix applications, and a cracked-and-broken worksheet.

Chapter 9: Establish Consistent Work and Many-Level Reviews (Step Eight). This chapter explains the methods to use to prevent future problems once a system or specific problem has been rectified. It proposes the use of check sheets to prevent conditions from slipping out of control. It also provides a mixed-part check sheet, which will serve as a guide for preventing future problems and managing problem-resolution activity. Last, it recommends that any important manufacturing responsibility be assigned to an advocate.

Chapter 10: Summary. This chapter ties everything together with a few handy lists.

There are also six appendixes, which contain more information on processes, the basis for decisions, various tests, more complicated examples, and definitions.

This book is full of examples of actual methods I have employed to determine solutions to problems. The discussion is keyed in to illustrations where possible, because they clarify the methods employed.

The book also shows you examples that you can apply to similar circumstances. It will increase your perceptions in dealing with manufacturing problems and, hopefully, aid you in becoming a manufacturing or service problem-solving expert.

I only wish that this information had been available in a course during my engineering or technical training; it could have saved me years of problem-solving effort.

CHAPTER 1

Define the Problem

Step One

This chapter starts with the most basic element in problem solving: defining the problem.

Some people might not understand the difference between the problem's definition and the problem's cause. It's important to understand this difference before you venture into the realm of problem solving.

The Cause and Why It Is Important

Consider for a minute a plastic part formed in a molding machine. The machine has at least two moveable parts that are joined before the resin is injected into the mold. One of these is the *cope* (the top) and the other is the *drag* (the bottom). These two sections must join flawlessly to produce components that don't have a substantial knit line (mismatch) where they meet. Mismatch is prevented by using mating pins in the drag section and bushings in the cope section. These components keep the pattern sections from shifting in the north, south, east, and westerly directions. As long as the pins and bushings are new, the mating can be almost flawless and imperceptible. However, when there is excessive wear on any of these control components, the pattern can shift significantly. When that happens, a visual knit line with the potential of a sharp fin will be produced.

Chapter 1 | Define the Problem

The *problem* is the manufacture of a part or process that contains a defect. The problem *definition* is that the part is unacceptable because it has a visible defect (knit line and a sharp fin). Put another way, the problem definition can be simply stated as, "The part is defective, because it won't fulfill its intended purpose." (If an O-ring had to be applied to this component, for example, it would fail because the fin would cut and damage the O-ring during operation.)

An ensuing investigation would seek to identify conditions that could influence the quality of the part under question. The conditions, or clues, are generated by comparing a properly manufactured part with a mismatched part.

The cause of the problem in this case is excessive wear on the pin and bushing components, which allowed the poor positioning of the mating pattern. This poor positioning allowed the mismatch and the unacceptable knit line. To eliminate the cause of the mismatch, it is necessary to change the worn pins and or bushings.

Hopefully, this distinction will become more apparent as you observe the samples and illustrations in the book. In the past, many manufacturers would recognize the mismatch and fin as being detrimental and would have assigned an individual to file down the unsatisfactory components to remove the mismatch and the fin. This is reactive and does not solve the problem or prevent it from happening again. The point of this book is to teach you how to identify a problem, remove the inadequacy, and prevent it from happening again.

So the difference between a problem definition and its cause might be expressed as follows. The problem definition is a description of the conditional state (it fails to meet a requirement) of the part that created or resulted in an inadequate product. You then augment that description by uncovering relevant clues.

The cause of the problem, on the other hand, is that condition to which you apply corrective steps in order to eliminate the observed detrimental characteristics. You focus on pinpointing the specific mechanism, practice, material, process, or environment that made the component characteristics unacceptable.

■ **Note** The cause of a problem is the specific mechanism, practice, material, process, or environment that made the component characteristics unacceptable.

Problems: Inherent in Manufacturing

All manufacturing problems have causes. And all causes are specific conditions that can eventually create customer discontent. Customer discontent is unpredictable, because it relates to users' expectations and values. In manufacturing, defects are inadequacies in the process that affect the final product quality. These inadequacies can affect aesthetics, smooth product flow, product function, or equipment downtime. They may even cause costly recalls. In service operations, defects may result from customers comparing high expectations to marginal results.

The good news is that what you learn from this book is applicable to simple and complex manufacturing or service problems.

Since each problem is the result of some recognized fault, you can use "cause identification" to focus on a specific flaw present within the problem. The term *main cause* describes the detrimental condition that prevents a desired outcome. You have identified it and referred to it as the root cause, the source of the problem, the main variable, or another identifier. To reduce ambiguity, it's called *the cause* from this point on.

The cause of a problem is not the same as the definition of the problem, as mentioned. It is the factor that aggravates a condition that allows a problem to form. For example, the sharp fin on a plastic part can create an unsafe handling condition. That is a safety problem. The characteristic of the safety problem is the fin. As you've seen, the misalignment of two mating mold sections can allow mismatch and cause sharp fins. This improper mating creates a gap for excess plastic to escape, which then forms the sharp fin.

In this case, you can prevent the safety problem only by correcting the attribute that caused it. You would have to improve the mold mating, or sealing surfaces, to prevent the sharp fin from forming. The sharp fin is therefore a *characteristic* of the safety problem, but it is not the *cause* of the problem.

The first step in problem solving is therefore to adequately and completely define the problem. It is imperative that you take the time and effort needed to define your problem thoroughly. The extra effort you take to define the problem thoroughly will enable you to identify the cause quickly. This results in untold timesaving dividends.

Note   The first step in solving a manufacturing problem is to adequately and completely define it. Don't jump to a solution before you have defined the problem, or you will probably have to solve it again in the near future.

Another example: An engine assembly line was experiencing a high reject rate due to a leak test. We defined the condition in this manner, "Engines are failing the leak test at the front water seal areas." We then conducted an evaluation to determine the characteristics responsible for causing the engine leak rejects.

The evaluation revealed that the front seal assemblies had traces of a foreign material (FM) that coated the outside diameter of the seal. The foreign material was trapped between the seal's outside diameter and the engine-housing wall. We removed and inspected many leaking seals to generate clues.

At this point, some people insisted that a blow-off (a high-compression jet of air directed on a surface to clean or dry it) be applied to each seal to remove the FM. But then we noticed that there was an unusual scratch (wear mark) on the seal's coating, which proved to be an important clue. Study showed that this scratch was due to seal ram misalignment with the engine assembly, which sheared the coating during installation. The misalignment of the positioning locators proved to be the cause.

The immediate problem was the rejection of engines due to a leak test failure. However, the cause of the problem was not the sheared traces of material that were found on the leaking engines. The cause of the problem was the misalignment of the positioning locators, which allowed the uneven shearing and enabled the particles to be trapped in the assembly. The trace of material found on the rejected units and the unusual installation wear pattern were characteristics of the leaking, and they were valuable clues in identifying the problem.

Conditions: Clues to Solving Problems

Each of the following conditions was a characteristic we found in different studies involving leaking seals in engines or other products:

- Misplaced or insufficient grease application
- Inadequate dust seal lubrication
- Foreign material on the mating surfaces
- Mismatch on the molded part
- Imperfect rubber coating
- Packaging components found in the assembly
- Reworked but unacceptable parts found in the assembly
- Strings and fibers negatively affecting the assembly
- Seals with dimples created during the assembly process

It should be apparent from this list that there are numerous conditions that can allow a seal problem to occur. Each condition contains a plethora of information that is not observable to the untrained eye. Each characteristic can be a valuable clue to identify and solve the puzzle. It is not uncommon to find a major clue when studying a defective part or the results from a troubled process.

Tip Always approach an industrial problem as if it were a crime scene. This enables you to glean as much information as possible and find more clues. Always check the "victims" for valuable clues.

Key to Gleaning Clues and Solving Problems

One of the tools for generating clues is a *problem corrective action worksheet* (Figure 1-1). You can use this sheet to verify acceptable and established practices and to assess compliance with current work instructions and procedures. The instructions and procedures are sometimes referred to as *consistent work instructions*. They specify the detailed manpower, tools, methods, procedures, equipment, and measuring devices used in a particular operation. In addition, environmental and required safety equipment might be specified, although these two concerns are not explained here.

Problem Corrective Action Worksheet

Problem Description: _____

Failure Mode: _____

Indicate Status:

Question	Current Operation YES	Current Operation NO	Question (Upstream)	Upstream YES	Upstream NO
Is Consistent Work specified on this Job?			Material is unchanged in last 3 months?		
Is Consistent work followed?			No Major Equipment repairs in last 3 months?		
Is Consistent work posted on site?			No special quality issues last 3 months?		
Are Jobs done on all shifts the same?			Same Tier 1 supplier used for last 3 months?		
Do Operators understand standards?			Same Tier 2 supplier used for last 3 months?		
Are Regular Operator on the job?			Management is unchanged in last 3 months?		
Are Operator properly trained?			Process has no Major shift in last 3 months?		
Do Operators understand job requirements?			Are all processes centered to Nominal?		
Do Operators know how to flag problems?			Does part receive a First Piece inspection?		
Are Operators instructed to report concerns?			Does a last piece Inspection close the loop?		
Is Job Error Proofed?			Is this first time this defect happened?		
Error proofing system is checked daily?			Defect occurrence is random not time related?		
Are Correct Tools and Fixtures on Job?			This defect can't be made on purpose?		
All tools present & functioning correctly?			Is there adequate lighting on the job?		
Does Welder operation have PM Schedule?			Is inspection area lighting sufficient?		
Is All PM work completed to schedule?			Is ancillary lighting available on the job?		
Is an Operator Check Sheet being used?			Is Cell concept used for this part?		
Are Correct parts being used?			Is Team Concept used for operation?		
Are Part routings current and followed?			Is a Quality Gate in Place for defect?		
Are Parts stocked in correct location?			Can a Photograph of defect be provided?		
Can Good and Bad parts be identified?			Is Statistical Process Control Used?		
Do Good & Bad parts look different?			Are all parts identified with tags for use?		
Are Rejected Parts placed into LOCKBOX?			Are all old ID tags removed before loading?		
Is there only one flow path for parts?			Are all gages up to calibration checks?		
Are parts in specification?			Is operator queried regarding the		
Problem is not caused by rework?			There are no components/parts on the floor?		
Rework is done on the line?			Reworked parts are identified and kept separate?		
Many Stage Checks are being performed?			No parts are found outside of the routing route?		
Component is unchanged in last 3 months?			Bad parts have dissimilar date codes?		
Operation(s) unchanged in last 3 months?			Bad parts come from different dies?		
Checks are made after set-up Adjustments?			Bad parts come from different locations?		
Variable gages are calibrated/used?			Tier 1 Supplier verifies that there is no change?		
Operation Date Codes used?			Tier 2 Supplier verifies that there is no change?		
Other?			Suppliers are required to certify their shipments?		
Other?			Suppliers have certified all their shipments?		

Notes: _____

Figure 1-1. The Problem Corrective Action Worksheet

The questions shown in the worksheet are self-explanatory, as they evaluate conditions in the affected and previous operations. Your first step should always be to evaluate these established benchmarks, in order to prevent wasted efforts. Answers that result in a "Yes" are acceptable. You need to study the answers that result in a "No" more fully—they are clues that can help you find the cause of problems.

You can correct the vast majority of industrial problems by using the problem corrective action worksheet. The clues you gather from using this sheet help to pinpoint systemic conditions, which are more easily corrected than more complex, obtuse problems. For more complicated problems, you will need to use all eight steps in the process.

If you don't find any "No" evaluations, other areas are suspect. You might need to check for missing baskets on a casting overhead in a foundry to prevent hotcrack scrap. Or, you might need to audit a chemical manufacturer's cooling tower temperatures to control a process. As you find conditions that cause detrimental effects, add them to your own corrective action worksheet.

Tip   Keep all worksheets as a starting point for each of your future evaluations.

In any case, it is advisable to request the work sheet information from your tier 1 and tier 2 suppliers[1] to ensure that they have not changed materials, components, operations, processes, equipments, or systems that you have certified. As an example, a supplier once changed the component raw material for a part sintering operation (creating objects from powders) from red iron oxide to black iron oxide without permission. This caused the product of the manufacturing process to fail.

Another supplier changed the approved oven temperature for curing foam molds without permission, and it resulted in warped crankshaft castings.

Both suppliers later insisted that they made these "minor" changes in order to save costs. Unfortunately, they denied making these changes until very late in the problem evaluation, which added cost and scheduling problems.

As a sad matter of fact, almost 40% of the problems we experienced with 200 suppliers over a period of one year were caused by unauthorized material changes, process changes, or the lack of consistent work items as identified on the problem

---

[1]A *tier 1 supplier* supplies goods directly to you, the customer. A *tier 2 supplier* supplies parts, services, or components to the tier 1 supplier, who then utilizes the materials or service and sends the product to you, the customer.

solving worksheet. Admittedly, not all problems are so easily solved. Sometimes scrap or defective components require in-depth study in order to identify the characteristics that lead to the discovery and elimination of the cause.

Once you have established consistent work procedures for your internal and external suppliers, you can focus on relevant clues.

Summary

This chapter introduced basic material that will allow you to progress more easily through the remaining chapters. You should now feel more confident about focusing on the problem and evaluating its characteristics. The next chapter deals with defect identification. It contains tables that will help you isolate important separation criteria identifiable within the problem. These tables will become valuable tools for identifying defects.

Since you are now able to create an adequate description of the problem and describe its physical conditions, it is time to move on to the next chapter.

CHAPTER 2

Define Fault Characteristics
Step Two

The second step in the problem-solving process is defining the fault characteristics. Some call it *defect identification*. Visual observation may provide clues that can point to conditions related to the problem, and can aid in generating clues. Dents on a bearing, for example, may signify that the forming process is askew.

Say, for example, that you produced a plastic component that is not up to blueprint specifications. You must discover and eliminate the cause of the discrepancy, which you'll do by generating clues. As you'll see, you can do this by comparing good and defective parts in order to identify relative characteristics. Characteristics that exist on most of the defective parts but not on the acceptable parts provide insight into the cause of the problem.

Defects: Clues to a Problem's Cause

This step is often overlooked when a problem develops. That is not unusual, because as you see often in your own life, many people rush to propose a solution without understanding the problem. An undefined problem can often lead to chaos. You could end up spending days discussing a problem only to discover that your investigators or team members have an entirely different interpretation of the problem. That's why it's important to agree on a problem's definition first.

Chapter 2 | Define Fault Characteristics

A manager leading an investigation can expedite solutions by focusing the team with a thorough understanding of the *flaw*,[1] the severity of the problem, and the location and the incidence of the fault.

It is necessary to achieve consensus from almost the entire team as to the true identity of a fault before analyzing a problem. This is not dissimilar from the methods used by some surgeons to identify the knee of the patient requiring replacement before starting a knee replacement operation. Some patients are savvy and prevent a mistake by writing "the other knee" on the good knee with a marking pen before their knee replacement operation. (Sad but true.) But the point is that everyone on the surgical team must agree on the knee that needs the operation (and be correct!).

■ **Note** Once you agree on the problem definition, you can start your search for clues, and that search starts with methods to reveal defects. You may even acquire enough clues to uncover a problem's cause.

Start with Conditional Data

How do you arrive at a consensus? To begin with, you and your team can collect conditional and quantifiable data. Conditional data provides an overall macroscopic view of the problem. When you visually compare good and defective samples that have varying rates of gas holes—known as *porosity*—you're collecting conditional data. Conditional data can also be the difference in a part path going through the manufacturing process. Quantifiable data, on the other hand, describes the number of defects or measurement values obtained.

Conditional data includes identifiers related to the process, such as manpower, suppliers, materials, methods, environment, and other conditions that may have contributed to the defect. These data spell out the process parameters that should be considered as part of the problem's cause prior to collecting quantifiable data. Table 2-1 lists important items when assessing conditional data.

---

[1] A *flaw* is a fault having any characteristic that detracts from the intended purpose of a process or service.

Table 2-1. Process Definition Table

Condition	Response
Is there only one flow path for the process?	
Is this operation done at only one station?	
Are multiple flow paths present and identified?	
Is the machining done on only one machine?	
Is the assembly operation done at only one station?	
Are variable gauges used to collect data?	
Are all parts tested on one test machine?	
Is there more than one serial or die used?	
Do all dies or serials have the same problem?	
Do other plants have problems with the same parts?	
Is the percentage defective generally consistent?	
Does the part meet the blueprint specification?	
Are the blueprint specifications ambiguous?	
Has the raw material been changed in last three months?	
Has the process been changed in last three months?	
Was there major equipment repair in last three months?	
Could this part be an off-line repair?	
Was a repaired part returned to the correct operation?	
Are reworked parts identified and kept separate?	
Can the failure be made on purpose?	
Have five good and five bad parts been captured?	
Have five good and five bad parts been measured?	
Is there clear separation of part differences?	
Other: Has the vendor certified his material and process?	

You must verify the information in Table 2-1; it must not be taken for granted. For example, the question, "Has the vendor certified his material and process?" can be revealing. Invariably, at least 50% of strength problems, for example, are due to unauthorized supplier changes to materials, processes, routing, handling, or reworking items without approval. As you'll read in a later chapter, for example, a supplier attempting to cut costs substituted red iron oxide for black iron oxide without permission—and with disastrous results.

Collect Quantifiable Data

Quantifiable data is more microscopic than conditional data. It provides actual measurements of the defect, the incidence of its presence, or some other means of assessment that allows you to make an evaluative comparison. If you can accurately describe and quantify the condition, you are more likely to understand the circumstances that generated it. And, if you can measure it, you can develop controls to maximize or minimize its presence. For example, a comparison of five good and five defective parts can be used to generate ratings or scores that can be used to determine which of two materials is better or if a process correction is acceptable.

You can use Table 2-2, which shows the "defect scene characteristics" list, to record features or traits that you observe about the defective part or process. Although this list of characteristics is not all-inclusive, it will provide a sound basis to start. Look for these characteristics during all your problem investigations.

Table 2-2. Defect Scene Characteristics: Are Any of These Present on the Parts?

Characteristic	Yes		Yes
Abuse		Fractures	
Bad Threads		Hardness Differences	
Bends		Looseness Differences	
Bows		Malformations	
Burrs		Non-Random Patterns	
Chatter		Partial Formation	
Chips/Slivers		Partial Machining	
Color Differences		Point of Origination	
Cracks		Scratches	
Creases		Shears	
Direction of Break		Shrinkage	
Discoloration		Size Differences	
Drops Involved		Smudges	
Evidence of Heat		Voids /Indentations	
Foreign Material		Witness Marks	
Other:		**Other:**	
Rework Identification		Interrupted Operation	
Incorrect Routing			
Mixed Parts			

The defect scene characteristics list captures some of the data available from the defect scene. Conditions identified from the defective part or service should be noted for evaluation. Abuse, breaks, scratches, witness or layout marks, discoloration, and other identifiers are contained on the list and provide the basis for asking follow-up questions. For example:

- What caused a mark?
- Where is the defect located?
- Does the defect have the same location on all the parts?
- Was it caused by an incident?
- Was the mark/damage due to the part shape or use?
- Was a steady force involved?
- Was an impact involved?
- Was a mating part marked or damaged?
- From which direction was the force applied?
- Was damage due to excessive force or was the part weak?
- Do all the defective parts contain the same mark?
- Do all defective parts contain identical marks of the same intensity?
- Do all defective parts have similar marks in the same relative area?
- Do good parts have similar marks as the defective parts?
- Where does the break/mark begin?
- Do other similar machines create the same mark?
- If the operation was interrupted, could the mark be a result?
- Could foreign material be included in the process causing the mark?

The defect scene characteristics list has been extremely useful in providing clues during difficult problem analysis. For example, recognizing that foam replicas of crankshafts used in a molding operation were warped led to the information that the supplier had changed the coating drying process without approval. This change caused the crankshafts to warp after the castings were poured.

It's equally important that you compare the good and defective parts. Record their differences using the "contrasts of two or more" list, as shown in Table 2-3.

Chapter 2 | Define Fault Characteristics

Table 2-3. Contrasts of Two or More Table: Where Do You Find Similarities or Differences?

Check	Yes	Yes
First Place to a Second Place	One Piece to Another Piece	
One Side to Another Side	One Event to Another Event	
One End to Another End	Within-Piece Changes	
One Surface to Another	Machine to Machine	
One Top to One Bottom	Line to Line	
Flatness to Flatness	Sample Duo to Duo	
Shape to Shape	Assembly to Assembly	
Smoothness to Another	Now to Then	
Good to Bad Group	Group to Group	
Individual to Individual	Batch to Batch	

Each clue provides a piece of the puzzle and leads to additional areas of inquiry. Not only do the conditions describe clues or defects, they also can be used to determine what caused the defects in the first place. You can compare individuals, duos,[2] characteristics, locations, patterns, statuses, or any other differences, as indicated in the list.

The "contrasts of two" exercise provides a way for you to focus on differences that you might not otherwise discern. If all of the defective parts have a characteristic that all of the good parts lack, you have enhanced the discovery process. For example, does the top half have more damage than the bottom half on each group? Or, if there are six different pattern serials, do they all have the same degree of the flaw in the same area?

Most anything can be compared in a contrast of two comparison. Two different machining flow paths could produce different dimensions, for example, as described in Appendix A. Or you might discover that a plate is warped when you're conducting a piece-to-piece comparison or a comparison for flatness. In most cases, the investigator will compare one sample to another and observe any differences.

---

[2] A *duo* is a matched comparison of two samples.

Industrial Problem Solving Simplified | 15

Summary

After reading this chapter, you should understand better how to capture and record relevant clues in the form of defects. It may be necessary to refocus your investigative team on the clues you generate. If it's not guided, the team will stray in a hunt for the solution.

Hopefully, the information in this chapter has provided you with an understanding of the methods that you can use to identify clues and defects. Defects might be the result of a faulty product or a failing process.

Remember, *problem definition* merely points to the failure of a component or system to meet requirements. In some cases, it may be an identified defect or characteristic of a component that makes it unacceptable. These defects are often clues to the *problem's cause,* which is the underlying condition that allows or creates the adverse or undesirable result.

Characteristics, such as defects, are the traits, qualities, or properties that are observable on a product or subject. You can evaluate them by comparing two or more parts, by looking at the variability generated by process differences, or by observing the defects visually. Scratches, gaps, discoloration, and so on, can be descriptive indicators of causal relationships that affect the problem. Other characteristics may be present but not significant. For example, some scuffing or mars may be an acceptable part of creating certain products.

The next chapter deals with a tool called a *concept diagram,* which is an important tool for finding and listing additional clues.

CHAPTER 3

Construct a Concept Sheet
Step Three

A basic understanding of the fault and the conditions that caused it are of paramount importance in understanding any type of defect. You must be able to change any unfavorable conditions that affect quality. So, as with all defects, determining what caused the flaw is important. Having focused on the fault in the previous chapters, it is now time to consider its cause.

Enter the Concept Sheet

Developing a *concept sheet* is the third step in formulating a proficient plan of attack for understanding the cause of a problem and solving it.

Each problem solver brings acquired knowledge to a study. Because of this, it's important to leverage the knowledge of the whole team when developing a method to solve a particular problem. Your problem-solving team's primary activity is to create a concept sheet. This sheet will help you determine which conditions are contributing to the problem.

A concept sheet defies precise definition. It can be a rudimentary sketch of a condition, or an intricate explanation of a process. A concept sheet is therefore anything that enables an observer to conceive, visualize, or create a theoretical model that simplifies cause identification. The weakness of developing concept sheets is the lack of innovation due to investigator inexperience. To help those lacking experience, and to aid those who need to hone their problem-solving skills, I provide many examples.

Chapter 3 | Construct a Concept Sheet

The following concept sheets were used to identify actual problems. Each sheet includes a brief explanation so you can apply it to related problems.

Poor Fusion Weld Concept Sheet

Problem: A heater hose assembly bracket was subject to broken welds before installation. Evaluation of the pipes showed there to be discoloration—signifying heat transfer—but no penetration at the weld-bead mating surfaces. The concept sheet we developed (see Figure 3-1) provided visual clues that allowed us to address inadequacies in the weld operation.

Solution: We realized it was necessary to change the weld direction from a vertical to a horizontal pass across the pipe and bracket interface to ensure quality welds. This new pattern was required because the process was not capable of ensuring proper flush mating between the components each time.

Industrial Problem Solving Simplified | 19

Poor Fusion Weld Concept Sheet

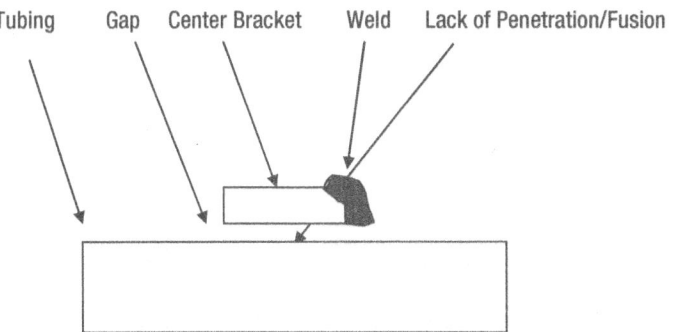

Causes of poor fusion of pipe/weld appears to be associated with the tip and gap:

1) Weld tip targeted wrong and concentrates heat on the center bracket.

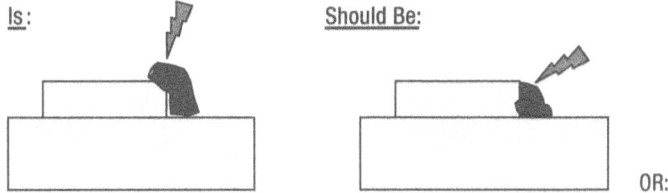

OR:

2) Gap is excessive and doesn't allow good contact and heat transfer.

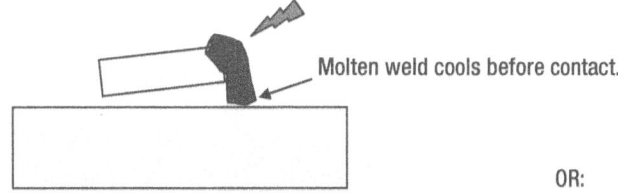

OR:

3) Fluid weld material cools before fusing with pipe.   OR:

4) Clamping of center bracket to pipe (FLUSH) before welding is inadequate.

Figure 3-1. Poor Fusion Weld Concept Sheet

Front Cover Oil Leak Concept Sheet

Problem: An engine front cover oil leak caused process rejects and affected first-time quality. We developed a concept sheet (see Figure 3-2) that facilitated the recognition of the possible causes. The sketch indicated that the fit in areas where the parts interfaced, damage, and the lack of a quality casting could all be creating an oil leak.

Concept Sheet

Problem: Front Cover Seal Leaks at Hot Test

Optimal Assembly:

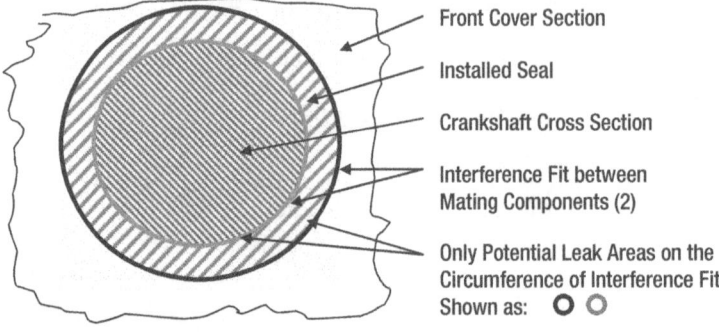

- Front Cover Section
- Installed Seal
- Crankshaft Cross Section
- Interference Fit between Mating Components (2)
- Only Potential Leak Areas on the Circumference of Interference Fit Shown as: ○ ○

Causes of Leak Failure:

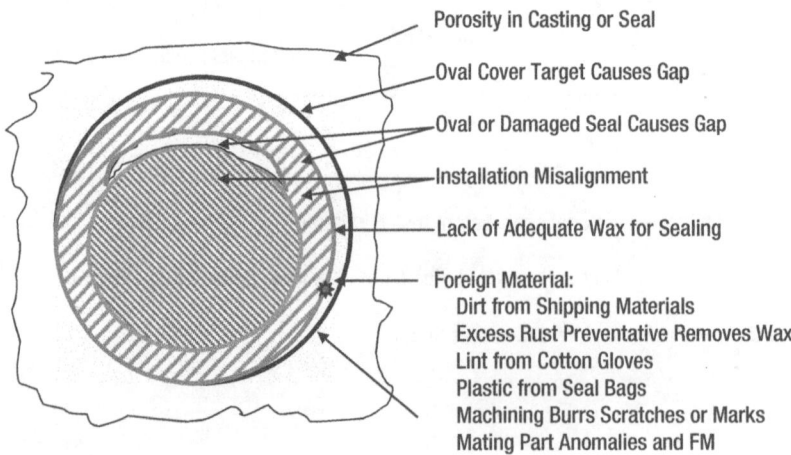

- Porosity in Casting or Seal
- Oval Cover Target Causes Gap
- Oval or Damaged Seal Causes Gap
- Installation Misalignment
- Lack of Adequate Wax for Sealing
- Foreign Material:
  - Dirt from Shipping Materials
  - Excess Rust Preventative Removes Wax
  - Lint from Cotton Gloves
  - Plastic from Seal Bags
  - Machining Burrs Scratches or Marks
  - Mating Part Anomalies and FM

Figure 3-2. Front Cover Oil Leak Concept Sheet

The sheet shows two sketches that were used for comparison. The top sketch indicates the optimal assembly of the components. We knew that the most significant leak areas were on the circumference of the mating components, which generally require a firm interference fit to ensure sealing.

The bottom sketch summarizes the problem conditions that we considered when attempting to define the possible leak paths. The concept sheet identifies the conditions we knew we should consider in addition to other visual inspections.

The leak conditions could include:

- Porosity in the casting
- Porosity in the seal
- Oval gap in the cover
- Oversize cover hole
- Oval or damaged seal
- Installation misalignment
- Machining burr, scratches, or marks
- Foreign material, such as dirt from packaging, lint or strings from gloves, packaging particles, or excess rust preventive
- Sealing lip damage

Solution: The leak was accompanied by pinhole porosity inherent in the casting adjacent to a boss area,[1] but 0.020 inches below the surface. Changing the boss location and the sectional area adjacent to the porosity-sensitive area eliminated the porosity. It appeared that the thick and thin metal sections solidified differently and voids were created when proper metal flow was not provided. Visual observation of the leak conditions was therefore rewarding.

Leaks Concept Sheet

What follows is a general template you can use to find leaks of all kinds, a common occurrence in most manufacturing facilities.

---

[1] *Boss* is a term used in molding operations to describe the knob-like presence of some solidified molded material that is intended to be machined and drilled to allow the attachment of components. It is a mounting point for fasteners.

Chapter 3 | Construct a Concept Sheet

Problem: All components containing castings, seals, gaskets, welded components, machined finishes, and so on, are subject to leaks that affect first-time quality. The concept sheet (see Figure 3-3) provides sketches of conditions as they should be, as well as how they may actually be. The difference between the two conditions provides clues that you can use to analyze the leak. Parts should mate, fit flush, and have the proper finish and orientation. When they don't—because of mismatches, voids, damage, poor finish, poor alignment, or positioning—leaks result.

Leaks Concept Sheet

First Rule: Ensure fault is thoroughly defined; (description, location, and severity)
Second Rule: Specify failure mode: provide photograph for each
Third Rule: Observe the flaw to determine clues of its origin
Fourth Rule: Identify: process materials, foreign material, chips, cotton gloves, strings, fuzz, cleaning solutions/chemicals, cardboard dust, rustproofing, etc.

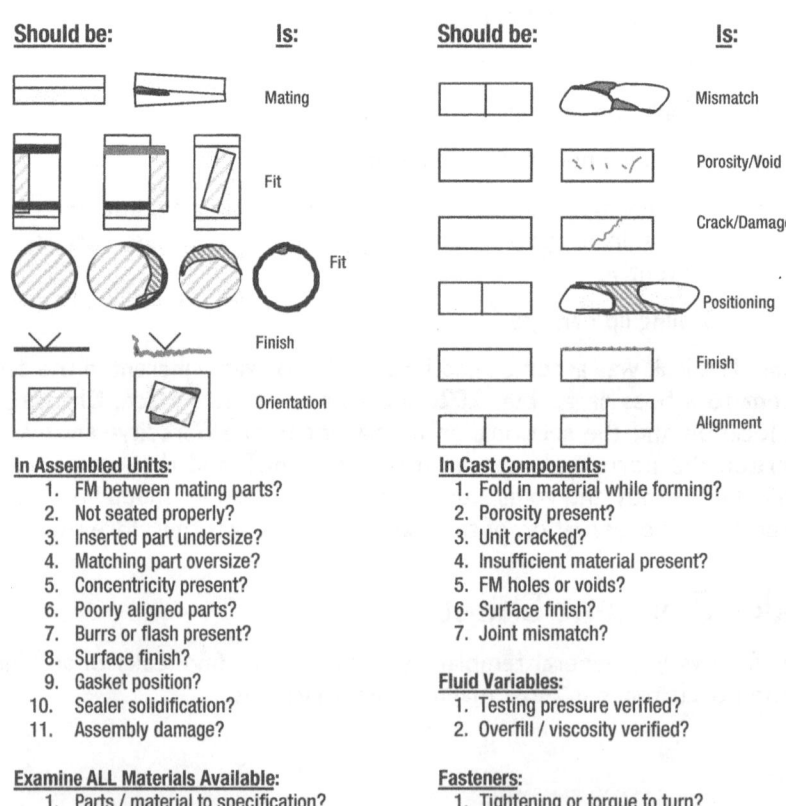

In Assembled Units:
1. FM between mating parts?
2. Not seated properly?
3. Inserted part undersize?
4. Matching part oversize?
5. Concentricity present?
6. Poorly aligned parts?
7. Burrs or flash present?
8. Surface finish?
9. Gasket position?
10. Sealer solidification?
11. Assembly damage?

Examine ALL Materials Available:
1. Parts / material to specification?
2. Witness marks and location?
3. Pattern serial number?
4. Sealing surface damaged?
5. Date code / Other ID
6. String / fuzz / FM originators?

In Cast Components:
1. Fold in material while forming?
2. Porosity present?
3. Unit cracked?
4. Insufficient material present?
5. FM holes or voids?
6. Surface finish?
7. Joint mismatch?

Fluid Variables:
1. Testing pressure verified?
2. Overfill / viscosity verified?

Fasteners:
1. Tightening or torque to turn?
2. Specified coating present?
3. Threads to size / form?
4. Fasteners stretched?
5. Parts to specification?
6. Excess coating or mispositioning?

Figure 3-3. General Leaks Concept Sheet

The top of the sheet also lists some useful rules that you can follow when you encounter leaks. The four rules are self-explanatory and will help you define the leak. First, you must ensure that the fault is properly defined. You then specify the failure mode[2] and capture it in a photograph if possible. Next, you identify potential origins of the leak; sometimes there are multiple causes. And, finally, you identify foreign material if it is present.

As you can see, the sketches are augmented by the clues presented below them. You should consider casting, assembly, fastener, and pressure variables. The lack of O-rings, misplaced gaskets, missing sealants, and damaged surfaces will become apparent with cursory examination.

In some cases, the problem may stem from an interaction. *Interactions* are when two things that have no individual effect adversely affect an operation in tandem. For example, if molten iron is spilled onto a dry floor during a foundry operation, it will splash and disperse. If the spill happens to fall on a water-soaked floor, an interaction is imminent. The heat from the molten iron changes the water to steam, which frees itself from the iron with explosive strength. This interaction can be violent.

In a later example, there is a condition where the application of the proper amount of grease provides an adequate seal if no major foreign material is present. It is even possible to have some foreign material present and still get an adequate seal. However, the presence of a small amount of foreign material may result in a leak if there is insufficient grease to provide a robust condition for sealing. This is also an interaction between variables.

Using the leak concept sheet assists you in identifying and solving most direct-effect leak causes. You might also be able to identify interaction components using this sheet, but you'll have to perform additional evaluations to be certain.

■ **Note** Concept sheets are applicable to all types of problems. You can express them as Venn or block diagrams and apply them to problems beyond traditional manufacturing, such as with sensing devices, electrical components, and many other areas.

---

[2]*Failure mode* is the description of the method of failure. A component might have failed because it is cracked, mismatched, or scratched, depending on the acceptance criteria used.

Chapter 3 | Construct a Concept Sheet

Failure Concept Sheet

The following explanation for a sensor failure that was rated as a "No Trouble Found" (NTF) by one investigator was later resolved under reexamination at the supplier. This case study uses an NTF concept sheet (see Figure 3-4) that you can use as an example for problems that initially don't seem to have a cause.

NTF (No Trouble Found) Failure Concept Sheet

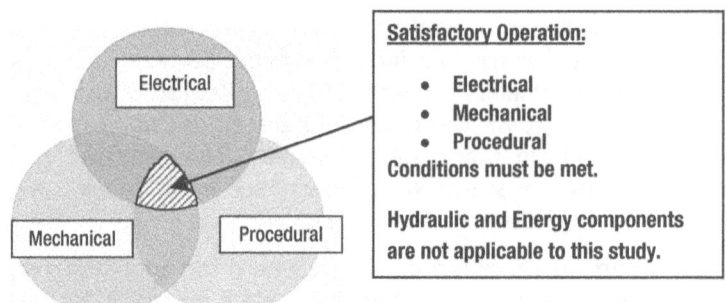

Potential Electrical Failure Modes (Most Likely Suspects):

Electrical:
Poor Electrical Contact	Internal Short	Induced Charge
Coil Pack Operation	Energy Supply	Static Charge
Broken Wire	Measurement Sensitivity	Ionization
Pinched Wire	Phantom Charge	

Mechanical:
Measurement Accuracy	O-Ring problem	Insertion Forces
Interference Fit	Burrs	Alignment
Assembly	Friction	Retainers
Sticking Ball	Nicks/Dings/Crack	Design
FM	Wrong Parts	Fatigue
Potential vs Kinetic Energy	Assembly Bad	Wear-out

Procedural:
Different Tests Used	Unauthorized Shortcuts	Misclassification
Different Parameters Checked	Data Interpretation	
Untrained Operator	Fluid Concentrations	
Improper Method	Process Changes	
Test Repeatability	Test Reliability	

Figure 3-4. No Trouble Found (NTF) Concept Sheet

Problem: Failed components at the assembly plant were found acceptable during reevaluation at the supplier location. The supplier suggested there was a phantom ionic charge that was generated in a sensing device when a tramp element (a contaminant) was allowed to interact with a transistor. The supplier further suggested that the sensor components had to be redesigned to shield the individual resistor components. This shielding was meant to eliminate the attraction between the highly negatively charged transistor and the positive ions generated by the raw material.

We found this supplier-proposed correction to be costly and unnecessary.

Naturally, the first thing we did to begin to solve the problem was to develop a concept sheet. The sketch on the NTF concept sheet in Figure 3-4 shows the relationship of the characteristics (electrical, mechanical, and procedural) that must be present in the installation of electrical and mechanical sensing devices. (Depending on your problem, hydraulic and energy components may also come in to play. They were not included on this diagram because they were not applicable to the study.) The overlap indicates that all three conditions must be present to allow satisfactory operation.

The table below the sketch identifies conditions that could affect each of the individual characteristics shown above. We considered the items shown in bold the most likely suspects before the investigation began. This table, however, did not restrict the study. Rather, it served to make the investigator more aware of the conditions that may have been present.

We knew that if any of the three input conditions were unsatisfactory, the sensor would not operate properly. As mentioned, since the sensor was acceptable at the supplier but failed upon installation, we had to return the failed sensor to the supplier for reevaluation. This required a costly rework operation in which the flawed component was removed and replaced.

We shipped the sensors that failed after installation overnight to the supplier for final analysis. The supplier retested the sensors on the test stand and found that they operated satisfactorily. The resulting report informed the customer that there was no trouble found (NTF) with the components. They were acceptable.

This type of result causes consternation, because it does not identify the problem and will result in recurrence of additional failures in the future. Upon pressing for a more definitive answer, the supplier reported a possible ion flow reason that proved to be incorrect. We needed to continue our study and so we did a full analysis, as shown in the parallel engineering study (see Figure 3-5).

Chapter 3 | Construct a Concept Sheet

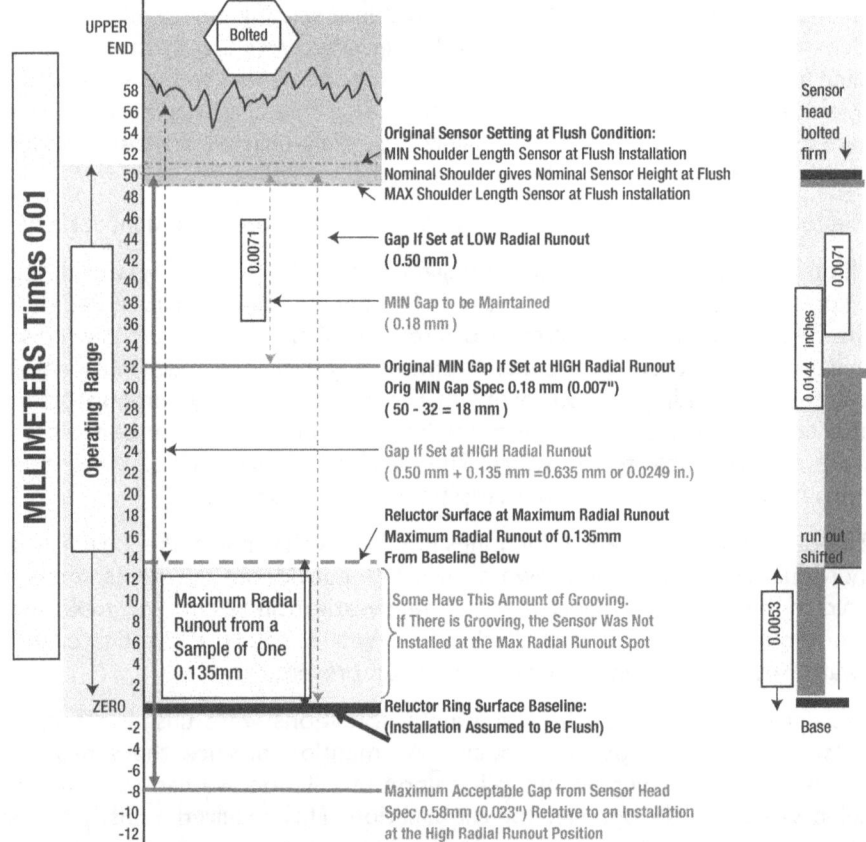

Figure 3-5. Crankshaft Sensor Operating Zone Analysis

Problem restated: Assembly failures were causing rework and were affecting quality because engines were failing tests during final testing. In one case, an important sensor that failed was positioned adjacent to the crankshaft within an engine assembly. Again, it was initially a case of NTF.

To start, we developed a concept sheet (see Figure 3-5). This analysis revealed that the sensor failures occurred only when the sensor was not inserted to the full depth. We found that the sensor could not be installed too close to the crankshaft. It could only be installed too far away from the crankshaft. If it was not installed fully, it could exceed the maximum sensing gap, causing failure.

As you can see, we sketched the sensor in position with a potential radial run-out condition to reduce the ambiguity. Once the sketch was made, we could visualize the bolted sensor and the interference forces that acted upon it. But while the sketch provided valuable clues, it did not provide conclusive evidence of the condition causing the malfunction.

Only after we created the operating zone analysis sketch did it become clear that it was not possible to violate the minimum gap. It was only possible to exceed the maximum gap if the sensor was not bolted tightly and flush to the crankshaft properly.

As a process comparison, look at Figure 3-6. Note the difference in skills and effort required to solve the sensor problem. It is sometimes faster and more convenient to compare the differences between the good and suspect parts.

Crankshaft Sensor Concept Example

Trace the condition described as "No Trouble Found" (NTF) on part retest

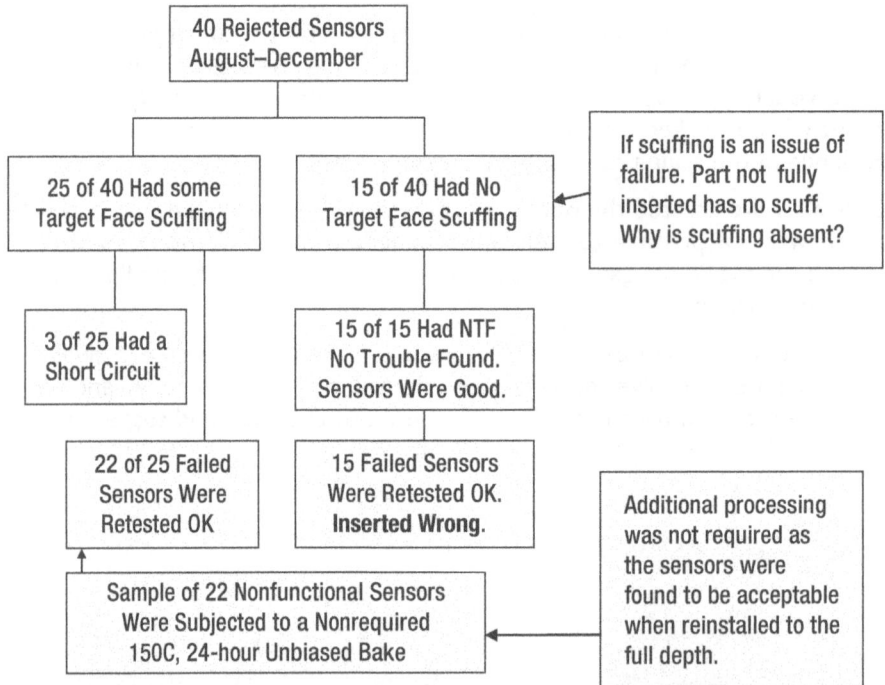

There are two types of failure present that can result in NTF retest values:
Sensors not inserted to proper depth are inoperable unless seated within the placement requirements tightly to established torque. (This can exist at assembly line only.)

Supplier indicated that there is an internal Ion flow or static charge that affects sensor operation. (This can exist at assembler and customer.) Ion presence is due to sodium from the wafer manufacturing process which is left inside the chip after manufacture.

<u>Recommendations:</u>
Assembler should ensure that sensor is always inserted tightly to proper depth upon installation.

<u>Supplier should:</u>
Make no changes at this time. Operator at engine assembly line was using an improper method and was being retrained in the correct installation procedure.

Figure 3-6. Crankshaft Sensor Analysis Block Diagram

There are two types of failure that can result in NTF retest values:

- Sensors not inserted to proper depth are inoperable unless seated tightly within the placement requirements to established torque. (This condition can exist at the assembly line only.)
- Supplier indicated that there is an internal ion flow or static charge that affects sensor operation. (This can exist either at the assembly line or with the customer.) Ion presence is due to sodium from the wafer-manufacturing process that is left inside the chip after manufacture.

Solution: We retrained assemblers to ensure that the sensor is always inserted tightly to the proper depth upon installation. In addition, we recommended that the supplier make no changes at this time. Operator at engine assembly line was using an improper method and was being retrained in the correct installation procedure.

Case note: The example in Figure 3-6 shows the summary of the study of the defective parts. Again, comparing conditions of good to bad parts proved useful in determining the problem. Since almost all of the rejected parts retested satisfactorily at the supplier, it was apparent that scuffing was not a cause of the problem. However, it was a useful clue.

The fact that a large proportion of the rejected parts had no scuffing was relevant, as it exhibited a visual clue to the installation operation. The failures were caused by an assembler at the engine assembly plant who failed to fully seat the sensor before tightening, as evidenced by the lack of full contact scuffing. The inattention to this task caused an excessive operating gap and continuation of a surface charge. After training, posting instructions, and auditing, the problem was prevented from recurring. We identified the clues to this condition via the concept block diagram—although the scuffing did not cause the problem, it provided a clue that led to its solution.

Initially, these examples may appear too complex or difficult to understand. However, as you attain more problem-solving experience, recognizing major differences and influences becomes easier. You'll find that you quickly gain experience after working on one or two problems.

Bent Ignition Coil Connector Pins

This section considers a different manufacturing problem, which uses a product flow plan and a mating part to correct a problem at a car assembly plant. Workers were unable to connect the car-wiring harness to the engine after inserting it into the chassis. The plant reported that the engines had bent pins that prevented electrical connection. Workers had to relocate the vehicle and engine to a rework station where the defective engine was repaired or a replacement engine was installed. This was a time-consuming and costly operation. This problem also created what is referred to as a "spill," because there were hundreds of engines in storage awaiting shipment and they had to be sequestered to prevent delivery to the patron plant. Each stored engine required inspection before shipment.

Problem: A vehicle assembly plant reported rework because of bent ignition coil connector pins. The only identification available at the time of defect detection was the engine serial number, which was not entirely traceable for each component or operation. Because the engine was already installed in a vehicle, the required repairs were disruptive and costly.

We created a list of the data from each of the different potential sources of the problem.

Available from the supplier:

- Different suppliers who provided the components
- Number of manufacturing cavities for each component
- Blueprint tolerances
- Percentage fallout (percentage of products that were defective)

Available from the engine plant:

- Conveyor saddle number used for engine assembly
- Engine identification number
- Six identical test stands used to check engine final test
- Damaged pin location

Available from the vehicle assembly plant:

- Engine identification number
- Damaged pin location
- Classification of failures
- Individual or group failures

The first thing we did was collect the information required to generate clues.

In most cases, variable data is preferred over attribute data because it is more descriptive and lends itself to statistical analysis. In this case, however, it was better to visually observe the process to determine if a precondition of pin misalignment was present (see Figure 3-7). The bent pin on the left came from test stand 3110, whereas the acceptable pins on the right came from test stands 3070, 3080, 3090, 3100, and 3120. The area requiring investigation, therefore, was test stand 3110.

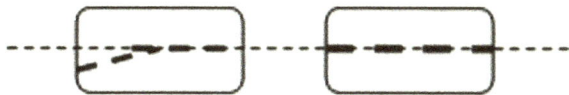

Figure 3-7. Bent Ignition Pin Drawing

A Focus on Process: Engine Assembly Routing

It's important to look at process flow. If you can follow the process and determine the point at which the component becomes damaged, you know something about which step is causing the defect. After some detective work, we discovered that the parts were acceptable at the engine plant but unacceptable at the assembly plant. Consequently, the damage was caused at the prior engine assembly or test operation, which was at our engine plant before the engine was shipped to the car assembler.

When we assembled components on the conveyor at the engine plant, the engines passed through one of six test stands before being shipped to the vehicle assembly plant. At the assembly plant, they are installed into vehicles and tested to ensure conformance with customer requirements.

We decided to evaluate if the problem found at the vehicle assembly plant was a result of engine plant problems or if it originated after leaving the engine plant. If components are found to be satisfactory at the end of a process, the problem is not with the previous operation(s). However, if any bent pins are found internally, the defect was caused internally and each previous related operation's output must be checked.

Because the bent pin condition could reasonably be the result of a twisted interference fit or an assembly problem, we acquired a matching component that was used in an assembly. This we used at the end of the engine plant manufacturing process to determine whether the collector pin connection could be made without difficulty. (A mating connector was used.)

A sampling of all connectors going down the line indicated difficulty in installing the connector on engines 146, 149, 151, and 153, which were all processed in sequence on test stand 3110. Since 4 of the 15 engines had a bent pin, and these engines were all tested on test stand 3110, that machine was shut down and scheduled for repairs. None of the other five test stands produced bent pins.

The clue that indicated which areas to investigate is shown in the Engine Routing example (see Figure 3-8) by the broken dotted line. Finding the cause requires an in-depth analysis of the station and the interacting components to ensure problem elimination. Without these clues, an investigation becomes excessive and time consuming.

Chapter 3 | Construct a Concept Sheet

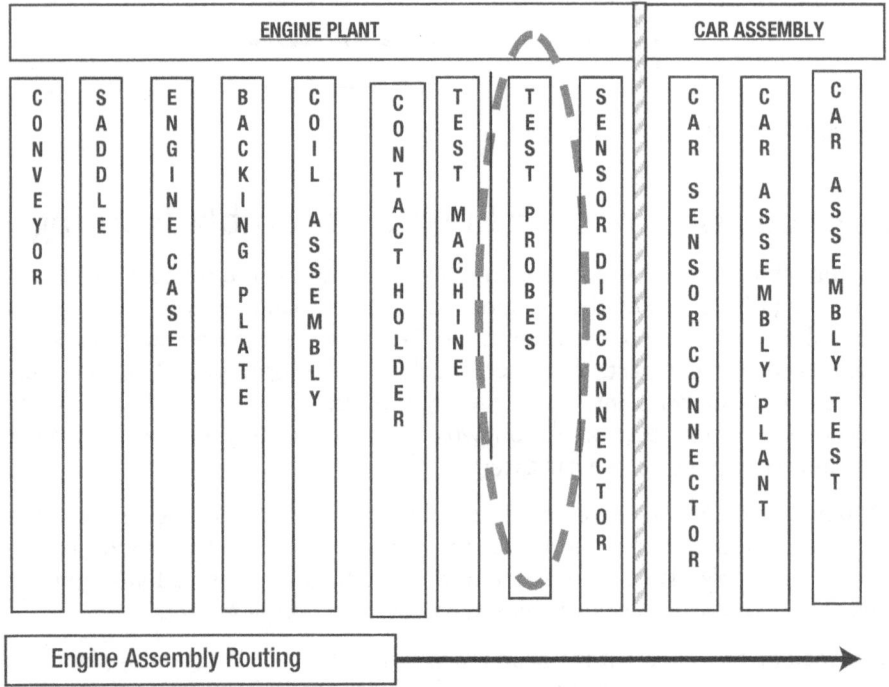

Figure 3-8. Engine Assembly Flow Path

Solution: A bracket with a broken bolt that held a probe in place was creating the bent pins. To ensure alignment with the connector on test stand 3110, we replaced the bolt-bracket assembly. The bolt was replaced in addition to the bracket, because the bracket's dimensional correctness was askew.

After making the repairs, the engine plant suffered no additional bent pin damage. The line was then audited with a matching connector component at the start and end of each shift. In addition, they checked the sequestered stock to ensure that no more defects would reach the customer.

The strongest clue leading to a solution was when we determined that all the failed assemblies came from the same test stand. However, replacing the bolt didn't mean the close of the project, because there was a reason that the bolt broke.

The assembly bracket required revision to allow proper seating of the bolt, and the damaged probe was replaced. One of the results of the bent pin problem was that the engine plant audited daily for bent pins, which prevented other malformed pins from being shipped to or reaching the assembly plant.

Industrial Problem Solving Simplified | 33

> **Note** These examples show how using a concept sheet in form of a diagram, routing sheet, or a sketch is important to problem analysis. The concept sheet allows the investigator to formulate the scenarios most likely to be present. You can use these scenarios to determine the most useful steps that can be used in a plan of attack.

PCB Board Process Map

Figure 3-9 shows a PCB process flow map, which is an outline of the routing process for the manufacture of an electrical PCB board. By starting at the upper-left corner and following the directional arrows as specified by the answers given to the questions, you can follow the progress and read the process controls and steps specified in the blocks on the right side of the sheet.

Summary

In every case, you have to address any problem you identify. You must put corrections in place to prevent recurrence. There must also be a plan to ensure that the problem will not resurface. Once you encounter a problem, it is prudent to assemble a file that can be used at a later date to provide information about the cause of the problem and the solution that you initiated. The lessons learned should also be reflected in changes to the design and process FMEA (failure mode effects analysis), as well as the control plan and job instructions for the process. These changes are made to ensure that the conditions that caused the problem are addressed in the design and quality planning for similar future products.

Using visual aids in generating clues for a plan of attack help you problem solve. This process includes evaluating the considerations and collecting the data, discussed in the following chapters. Before moving to data collection, however, I will discuss evaluation considerations for your plan of attack. You can evaluate these scenarios in order to determine which one provides the most influence and can be used in your plan of attack.

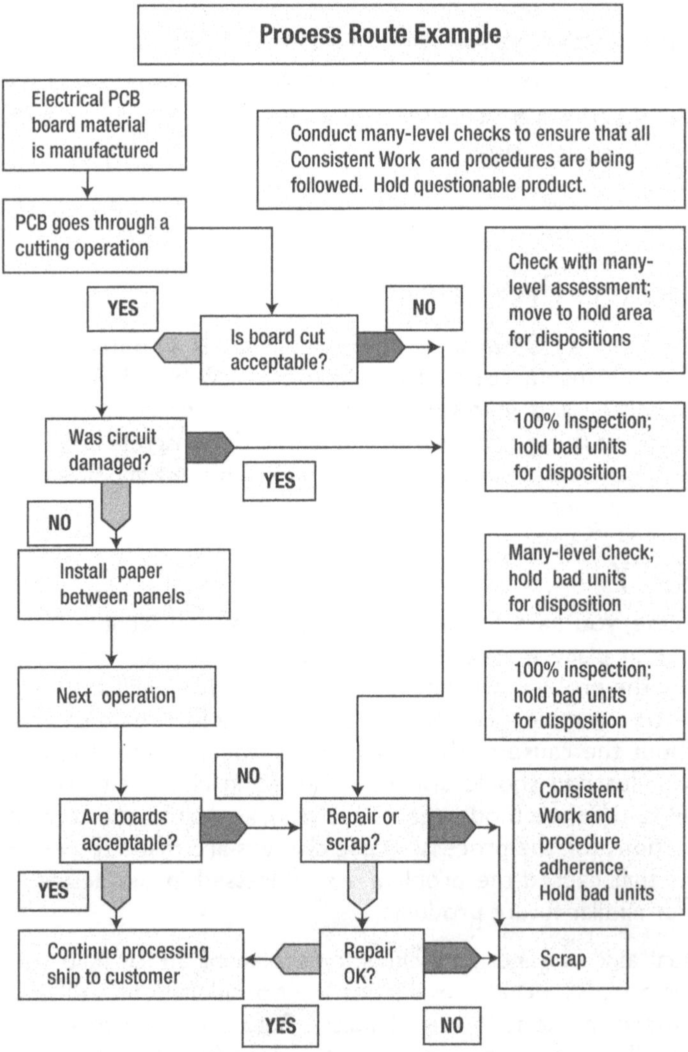

Chapter 3 | Construct a Concept Sheet

Figure 3-9. PCB Process Flow Example

CHAPTER 4

Develop a Plan of Attack

Step Four

The fourth step in the problem-solving process is developing a plan of attack based on observations, measurements, insights, experience, testing, and any other method that has provided clues. A plan of attack gives you a framework for solving a problem that focuses on the steps you should take based on the nature of the problem. As you'll see, it also includes assigning an advocate to critical areas.

You formulate the plan of attack after identifying the clues. Isolating the differences allows you to compare the best versus the worst conditions you observe. The description of the differences should be as detailed as possible; this detail will prevent you from unnecessarily evaluating unimportant conditions later.

Use Your Concept Sheet

At this stage, the concept sheet is very important in tying together the clues you have generated from the worksheets, process list, defect scene characteristics list, contrasts of two list, process routings, and other tools you used during problem analysis. The insights you gain from the concept sheet will directly influence your plan of attack.

For example, visual inspection of a defective weld indicated that there was not sufficient penetration of the mating components. We determined this by comparing five good and five rejected assemblies. We observed that all the job instructions were satisfactory and in effect. The welding head was not interfacing adequately with the components to be welded. Consequently, the plan of attack included comparing the samples, checking the compliance to job instructions, visually checking the process, determining which clues were relevant, and concluding what created the problem.

We adjusted the welding head, which resulted in proper weld depth penetration and assembly strength. We then compared five previously welded samples from the old weld operation with five from the new weld operation with the head repositioned. The latest welds were superior, and the change was verified.

The following paragraphs include some additional factors that may be of use when you're defining a more complex problem and determining its cause.

Note A *plan of attack* is the method conceived and established by the evaluators to move toward the goal of eliminating the flaw under study. It can involve direct observation, measurement, testing, or sending a component out for chemical analysis, among other things. The plan of attack is based on the complexity of the problem and the required analysis. Examples of problems you'd develop a plan of attack for: Why is the balloon bladder stretched? Why did the casting crankshaft break? Why do refractory bricks fall off of the space shuttle? Any action specified or directed that results in achieving the goal can be construed as a plan of attack. There is no single formula or template for devising a plan of attack. Each problem is different and requires different actions to arrive at a solution. What's more, there are different routes for solving the same problem. Two different teams might have two entirely different plans of attack for eliminating a problem.

A Simple Plan of Attack

Say, for example, that you have a part that does not fit. Generally, the first step is to determine whether the problem originates from the geometry of the part or from the forces that act upon it. When dealing with a part that does not fit, there are three checks you can make:

- Is this the part that should be installed here?
- Has the part been damaged?
- Is the part up to blueprint expectations?

If the problem is related to a broken or cracked part, there are also a few questions to be answered:

- Was the part too weak to handle the applied force?
- Was excessive force applied to acceptable parts?
- Was the damage the result of an impact or pressure force failure?

Comparing major differences between parts allows you to identify the strategy—the plan of attack—that will be the most fruitful.

You can determine any major differences by physically comparing five good and five bad parts. In general, the greater the observed differences between the two groups in malformation, scratches, dimensions, or other characteristics, the easier it is to see the potential problems. Needless to say, it is important to compare the same characteristics in each group. It's not helpful to compare a size mismatch in one group to surface chips in another.

Use the first three steps of problem resolution previously discussed to generate clues as to your next steps for your plan of attack. Observe how data collection can pinpoint the specific source of the problem. The evaluator can then assess the results to determine a path to resolution.

These points will be discussed more fully when you come to the cause correction sheet. You can compare differences between many different elements—sources, locations, lines, machines, processes, shifts, piece comparisons, locations within the piece, differences over time, and any other elements defined by the investigator.

Evaluation Considerations

Evaluation considerations are the criteria present in the manufacturing system because of the nature of the problem: Why is the diameter too small? Why did the lathe overheat? Why did the chemicals in the process congeal and solidify? You must reflect on these properties and evaluate possible causes that could affect the problem under study. Such causes could range from using the wrong drill, using the wrong viscosity lubrication oil, or achieving a critical temperature that has an effect on the process. The problem under study will determine which process characteristics are suspect or viable.

Evaluation is all about questioning. For example, why does a system make some acceptable parts and some that aren't acceptable? Process parameters have a direct, indirect, or an interactive effect on the quality of the parts being made. A drill with too small a diameter, for example, will result in an undersized hole (direct effect). A lathe that is overheating due to a lack of lubricant may result in shifting the machine ways that produce a lack of machined concentricity (indirect effect). Hot-testing components with a combustible foreign material coating can result in a fire or flame flash (interaction effect). Each of these can influence the final product.

Consequently, it is necessary to verify that the obvious problem sources lead to the actual problem. Further, you must prove that corrective changes will prevent the flaw from recurring. These diverse effects influence forming, machining, and assembly operations; each can influence process variation differences. Service outputs are also affected by process variations.

Photographs of Seal Conditions

Figures 4-1 and 4-2 provide examples of evaluation considerations.

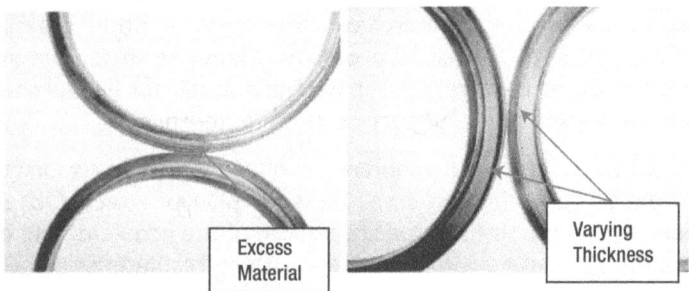

Figure 4-1. Photograph of Seal Conditions

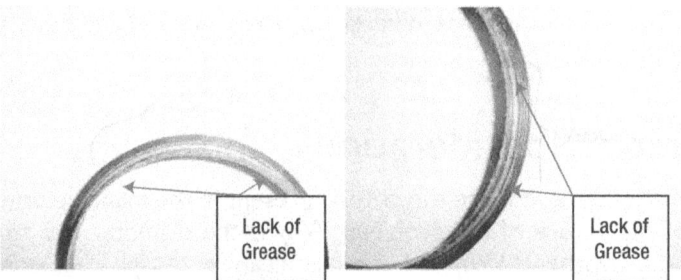

Figure 4-2. Illustration of Seal Grease Coverage

Engines were leaking oil and caused a problem requiring rework. (*Rework* is the process whereby a defective part or assembly is repaired or corrected to make it acceptable.) The seals were found to be the cause of the leaking oil, because the required coating of rubber wasn't uniformly dispersed along the periphery of the seals. In addition, some seals had an inadequate amount of grease applied to the sealing surfaces, which prevented accurate sealing. Here's what we found.

Excess material in the left photograph is shown as discoloration on the seals. A side-by-side comparison of a good seal and a bad seal showed that the material was not of the same thickness. This led to the observation that the excess material thickness was not the same during different cycles of the forming process. Nor was the leakage the same on different pattern serials.[1]

---

[1] A *pattern serial* is one of several identical forms located on a master pattern. A pattern is usually metal and provides a definite form or impression when forced into a retaining medium like moist sand. Some patterns contain six or more serials, which may be identical and are intended to all form identical cavities in the forming medium.

Since the material was of different thickness on the two parts that were made from different pattern locations and serials, there was something different in the simultaneous process (uneven coating). This indicated that the difference in the process was due to leakage while the rubber portion of the seal was being applied to the seal body.

Upon further inspection, when comparing five acceptable seals to five unacceptable seals, we found that the acceptable and unacceptable samples had dissimilar seal coatings and came from two different die cavities used in production. This is a very strong clue, because there was clear separation of the quality data (differences in the observable quality of the seals) between acceptable seals and defective ones. Some of the best pieces also had some foreign material, but not to the same extent as the worst pieces. (Always compare five of the best pieces to five of the worst pieces to see if there is a clear difference in the desired results. Comparing good vs. bad parts is a powerful diagnostic tool that will be explored in later sections.)

These clues enabled us to form a plan of attack. We were able to identify the supplier's pattern serials that were worn and defective; we speculated that they were not providing adequate sealing. In this case, it was necessary to physically measure the wear on the patterns and to conduct tests to eliminate the adverse conditions. Once they were repaired or replaced, the pattern serials made the necessary contact to prevent rubber leakage and formed the rubber seal section to specifications.

Simple Tools Work Well

The lack of sophisticated diagnostic equipment to provide thermal imaging, vibration analysis, or simulation should not be an insurmountable roadblock to solving problems. Simple tools—like micrometers, scales, and visual comparison ratings—are readily available to identify, isolate, test, and measure data. Of course, all the instruments and tools you use for measurement must be accurate and repeatable (calibrated).

Lack of specificity also creates a problem when fault dilemmas occur. There is too much time wasted in search of technical solutions that are applied with a "shotgun" approach. An approach that kills everything in the hopes that the cause will be permanently eliminated is inefficient and wasteful.

If a problem is not defined and agreed upon in its earliest stages, when the fault is recognized, it will create confusion in the ensuing investigation. Some people involved in problem solving will want to run useless tests that are not applicable. We had a fastener breakage problem that we solved by investigating marks appearing on the failed fasteners but not on the acceptable ones. These marks were caused by what is called a "double strike," in which the machine cycled twice on one part. You wouldn't believe the complicated tests and experiments proposed by people involved who had no concept of the problem. All of their proposals were impractical. The simpler the tool you use to investigate and achieve results, the faster you can obtain the solution.

One mistaken idea is that all identified variables must be adjusted even when they haven't been proven to affect the outcome. One manufacturer suggested that there were 25 different items that caused a contamination problem because he was able to list them on a cause-and-effect diagram. Actually, as it turned out, only one condition caused the problem.

Unless problems are sufficiently defined, there is a tendency—such as in this case—to concentrate efforts on obtaining information that is not pertinent or required.

The sketch in Figure 4-3 shows a side-by-side comparison of good versus bad parts. This comparison enables you to generate acceptable clues for problem solutions. This is another example where the basic practice of conducting a visual observation and constructing a concept sheet or sketch can help you generate a plan of attack.

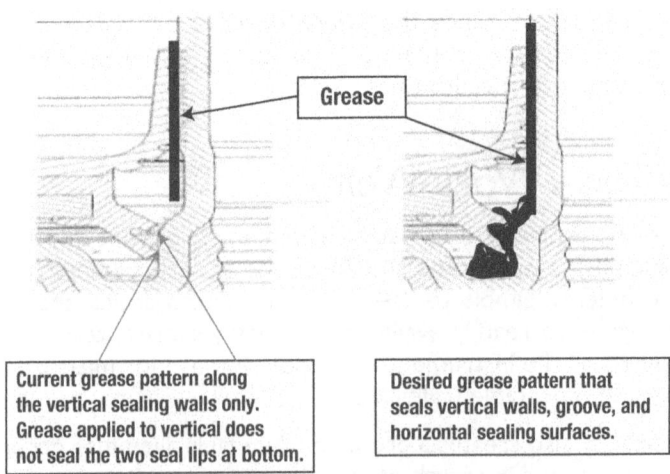

Figure 4-3. Comparison of Good and Bad Grease Application on the Seal

A seal leaking on a different engine caused line rejects and required rework. We started an investigation and the supplier was notified as part of a permanent step in the plant's plan of attack. (You can create these permanent steps over time by recognizing steps that are present in repetitive problems—like supplier involvement in creating problems—that you encounter.)

In this case, we observed foreign material on the seals and differences in the amount of grease present on each seal. Most seals from the five acceptable samples had more grease than the five leaking seal samples. The sketches in Figure 4-3 show the undesirable and desirable grease coatings. They were important clues. We created the sketch as part of the plan of attack in order to define the condition for all those concerned. We directed the supplier to observe the conditions at his packing facility as another step in the plan of attack, and he identified and eliminated the source of the foreign material. That's why notifying the supplier should always be included in your plan of attack.

Comparing the good and bad results should also be included in any plan of attack. Although each problem will have different clues and requirements, you should follow each of the steps previously discussed and incorporate them into your plan of attack. In addition to the differences in the amount of foreign material that was present on the seals from the different cavities, there was also a significant difference in the amount of grease applied to the seals before assembly.

One of the conditions, grease coating, was found to be statistically significant, and it was not included on the cause-and-effect diagram (not shown). Another suspect was contamination, which was caused by the pattern mismatch in the vulcanization process and resulted in a defective material flow called "bleeding." We decided to rate and compare the amount of contamination on the individual samples in order to gain a picture of the length and severity of the bleeding. We also measured the grease coating for acceptable or inadequate presence.

Figure 4-4 shows a plot of four quadrants, each experiencing acceptable or unacceptable grease application and the degree of contamination. The foreign material, FM, prevented flush sealing and allowed a potential leak path to exist due to an imperfect grease seal in the part interface area. Figure 4-4 shows this interaction.

Chapter 4 | Develop a Plan of Attack

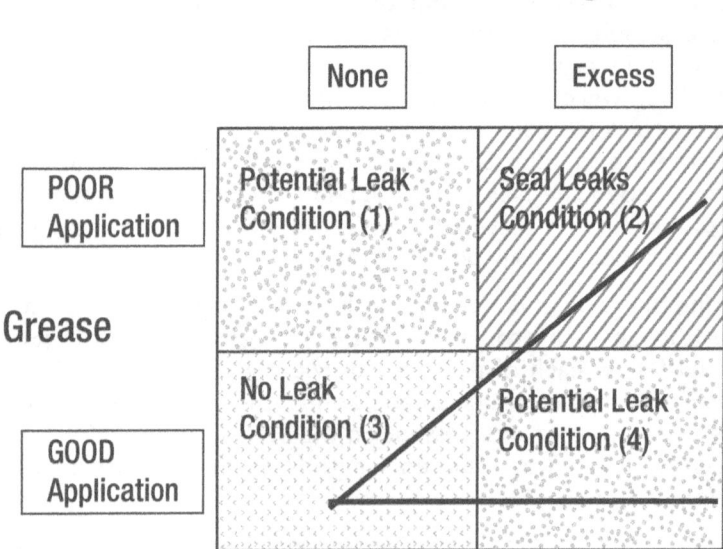

Figure 4-4. Interaction Diagram (Real Seal Air Test Leak)

This illustration is a simple way to visually represent the leaking seal problem condition. It consists of comparing results from two distinct variables—grease presence and the amount of rubber bleeding. It shows the following information in visual form:

- When there is no rubber bleeding on the seal, and it has a poor grease application, the seal could potentially leak.
- When there is excess bleeding and poor grease application, the seal leaks.
- When there is no bleeding and good grease application, there is no leak.
- When there is no rubber bleeding on the seal, and a poor grease application, the seal could potentially leak.
- There is a potential to leak when no bleeding is present.
- There is a potential to leak when grease application is acceptable.
- Leaks occur mainly when there is excess bleeding material and a poor grease application.

It should be clear now that you can use simple tools effectively to develop a plan of attack. Each of the previous steps discussed can be a specified step in the plan of attack. The way you attempt to generate clues—observing the job and checking the systems being used, comparing the good and bad samples, taking measurements, collecting data, constructing the concept sheets, creating sketches, and performing checks—should be included in a plan of attack with the individual problem requirements.

This section also introduced the concept of *interaction*, which is depicted in Figure 4-4 to illustrate a model that can be used with simple data to glean understanding of more difficult problems. For a full explanation and examples of what constitutes an interaction, please refer to Appendix B: Interaction Explained. The presence of interactive effects can also be studied and added to the plan of attack if it is appropriate to the problem.

Acceptable and Unacceptable Thrust Bearings

Your comparisons should contain at least five good and five bad examples of each condition, and they should be grouped side by side. (However, because of space limitations, only three are shown here.) Can you see in Figure 4-5 anything that might cause a machine jam?

Chapter 4 | Develop a Plan of Attack

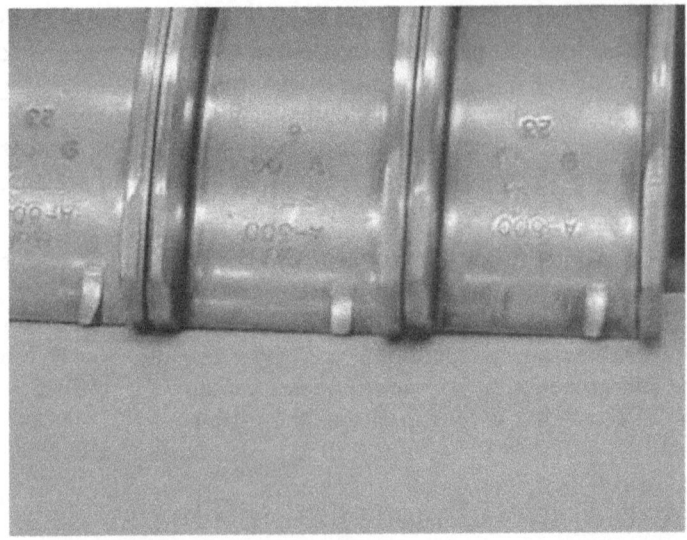

Figure 4-5. Photographs of the Same Three Thrust Bearings Taken from Different Angles

One good and two bad parts are shown here. When you compare parts side by side, it's easier to spot any differences. Two of the thrust bearings caused jams on the engine assembly line. Sometimes the quickest way to solve production problems is to collect a sample of the parts that are having difficulty and compare them to parts from a previous shipment or a previous trouble-free production run. As with any evaluation, compare the most acceptable samples to the most unacceptable samples available.

Industrial Problem Solving Simplified

Don't assume that a single sample can accurately show all the conditions. Using a single data point, no matter how clear it seems, will result in poor results unless it can be statistically defined and tested.

For example, team members might look at a single sample and have unbridled confidence about its relationship to the problem. In some cases, the data point may appear to be rational. In most cases, it will be proven to be groundless. Unfortunately, the one data point does not include any variability of the performance range of the population from which it came.

Generally this opinion is based on assumption and not on fact.

You can eliminate most uncertainty by comparing at least five samples of good and bad parts. The collection, analysis, and presentation of relevant information are keys to effective problem solving. You must identify relevant variables in order to allow efficient information gathering. (The basis for the choice of five good and five bad samples is explained in Appendix E.)

Now that you've pondered the evaluation considerations, it is time to consider the means for collecting data.

Let's go back to the thrust bearings. As mentioned, your comparison should contain at least five good and five bad samples of each condition, and they should be grouped by condition. Can you see anything that is different among the tabs in Figures 4-6 and 4-7 that would cause a machine jam? Some tabs have an excessive bend and a ding mark.

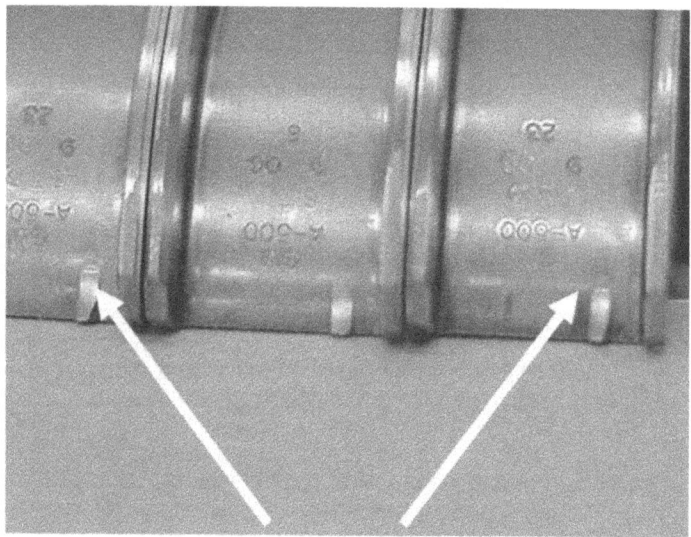

Figure 4-6. Dings in the Tabs Are Signs of Misalignment or Excess Penetration

Chapter 4 | Develop a Plan of Attack

Figure 4-7. The Excess Penetration Caused Interference in the Assembly

DON'T JUMP TO CONCLUSIONS

Imagine two investigators discussing a problem. The first one says that a new intake manifold casting has unacceptable porosity, and he thinks that it is caused by the people pouring the metal. The second person says they should determine the location of the unacceptable porosity.

The first investigator says that this must be the old problem that happened months ago. The solution then was to remove the skimming bars from the area where the iron is poured. So in his view, that's what they should do now. (Damp skimming bars were used in the past, which allowed moisture to be induced into the castings, thereby creating the porosity.)

He was correct that this condition was rectified in the past using this solution. However, it was not the cause of the problem this time. An inspection of the castings that were being run for the first time with a new pattern revealed an insufficient boss area that prevented continual liquid flow to the interior parts of the casting until they had solidified.

Jumping to a conclusion and a solution based on a sample of one is intolerable. It is always necessary to define the problem and its potential as the relative cause before applying solutions.

Here's where the plan of attack proves its value. It is a dependent process that focuses on the steps to be taken by the investigator depending upon the problem requirements. It can prevent a shotgun approach, where much time is wasted in search of irrelevant information. It helps you to avoid investigating unimportant areas or accepting conclusions based on a single sample.

Prevention

We would be negligent if we failed to discuss the concept of *prevention* within the plan of attack. The best way to solve quality problems is to prevent them before their inception. This section provides plans for the following (Figures 4-8 through 4-17):

- Plan for employee training
- Preventing mixed parts
- Preventing label problems
- Preventing missing operations
- Preventing foreign material
- Preserving part coatings
- Ensuring part identification
- Dealing with spills and sequestrations
- Ensuring communications
- Ensuring supplier compliance

These example plans that follow are preliminary and can aid in solving manufacturing quality problems. You can revise them to fit each manufacturer's plan and adjust them as you find additional areas of concern.

There is one key way to build prevention into a manufacturing system: assign responsibility. Responsibilities that are not assigned are generally ignored. Take for example the job of ensuring that the supplier will not send any mixed parts that could jam the customer's automatic material handling systems. The supervisor is responsible for instructing and checking his employees and their operations to ensure that the parts are not mixed. He might also be involved in checking a list of items in his area to discover and correct mixed parts and other manufacturing problems. But the supervisor's oversight should not be the only way to avoid the mixed-part problem.

The supervisor may suffer the brunt of mixed-part complaints, but the system may be set up to invite the problem. That's why, when managers are making plans for production, or recognizing corrections, they must ask questions:

- Who acts as the advocate in a quality-planning meeting or the (DFMEA or PFMEA) design and production failure mode analysis meetings?
- Who ensures that the job instructions are acceptable?
- Who ensures that effective employee training has been completed?

Chapter 4 | Develop a Plan of Attack

- Who ensures that checks are being made?
- Who should establish some method of recording and capturing significant items and practices that can prevent mixed parts from happening a second time for the same reason as the first?

As you can see, there should be many people making sure mixed parts do not get shipped. Allocating an advocate assigns responsibility and allows for accountability.

One of the ways to instill a prevention culture is to design simple plans that point out some of the faults that may occur during the manufacturing operation. The design doesn't have to be difficult or intricate. Rather, it is possible to create a system that recognizes certain failures or quality issues and provides questions that you can use to direct quality corrective actions as they are needed.

This may appear to be more difficult than it really is. If your facility does not currently have established quality planning meetings, plans, procedures, assigned responsibilities, training, actions, and follow-ups, you can use the following examples to create an initial system.

But first, let's briefly review what each of these items requires. *Quality planning meetings.* These meetings are called to address the quality considerations for the product or service being established. It could consist of discussions of design considerations (DFMEA), process considerations (PFMEA),[2] means to control the process (the control plan),[3] and procedures (safety, mixed parts, labels, missing operations, foreign material, coatings, part identification, spills and sequestrations, communications, supplier compliance, and notifications).

- *Control Plan.* The individual plan or outline discussed in the quality-planning meeting; it ensures that the subject of interest (such as a spill) is discussed.

---

[2]DFMEA stands for Design Failure Mode and Effects Analysis; PFMEA stands for Process Failure Mode and Effects Analysis. These are meetings attended by the respective department representatives to determine a plan to control the quality aspects of any problems. They differ for design, manufacturing, service, and other individual topics that are relative to the product.

[3]The *control plan* explains the specific method to organize and manage a manufacturing or service process. It also establishes checks and balances to ensure that problem conditions will have controls established and in place to prevent their occurrence. *Procedures and job instructions* establish consistent work practices and are generally written instructions on how to accomplish tasks required to perform the work.

- *Procedures.* There are about seven basic steps contained in each plan that follows. Each of the steps requires an assignee who will be responsible for ensuring that that step has been completed before the process begins.

- *Responsibility.* In addition, an alternate should be appointed and trained, so that there is continuity if the main assignee is absent due to promotion, illness, transfer, or layoff. This is necessary because when no one is assigned to a task, it isn't accomplished. If everyone is assigned to that same task, it's still likely to be ignored because no one is responsible for it. Each of the steps for each plan can be assigned to one or more individuals. However, it's more efficient when there is one assigned, responsible individual for each of the plans and there is one assigned, responsible alternate for that entire plan.

- *Training.* After ensuring that the part or process is discussed in the DFMEA and PFMEA and the procedures have been established, it's time to conduct the training. If you've conducted each of these considerations satisfactorily, you can check the operation to see if it is operating properly. If the process has not been in compliance with each of the preceding steps, the personnel in the quality-planning meeting must reevaluate each question or decision point that is not in compliance to what is desired. (This can be visualized as shown by the "no" arrow designations as shown in Figures 4-8).

- *Actions and follow-up.* Once you complete all the preliminary steps, you can start the process. You should check the start-up process and the ensuing operations at many levels, and on a randomized basis at specified intervals. If any deviation is observed, the assignee (or alternate) should respond and provide actions to prevent recurrence.

All corrective actions should be initiated as quickly as possible and should be incorporated into the DFMEA, the PFMEA, the control plan, the procedures, and the checks. This helps ensure that, once the problems are corrected, they won't recur.

There are ten preliminary plans provided in the ensuing pages. Each plan represents a simple method that you can use to improve the current process if one is not available at your facility.

Chapter 4 | Develop a Plan of Attack

There are other considerations that you can develop later, as required. These additions include:

- First piece inspections
- Routing parts
- Returned materials handling
- Container cleanliness
- Tagging suspect materials
- Others to be determined

Each of the following ten plans is self-explanatory. If you ask the question and attain a positive answer, continue to the next lower level. When you arrive at a "no" answer, you must take immediate corrective actions.

Now that you have the basic concept, additional plans can be developed. Each time you find a deviation in the system, correct it and adjust the process to improve the internal or outgoing quality level.

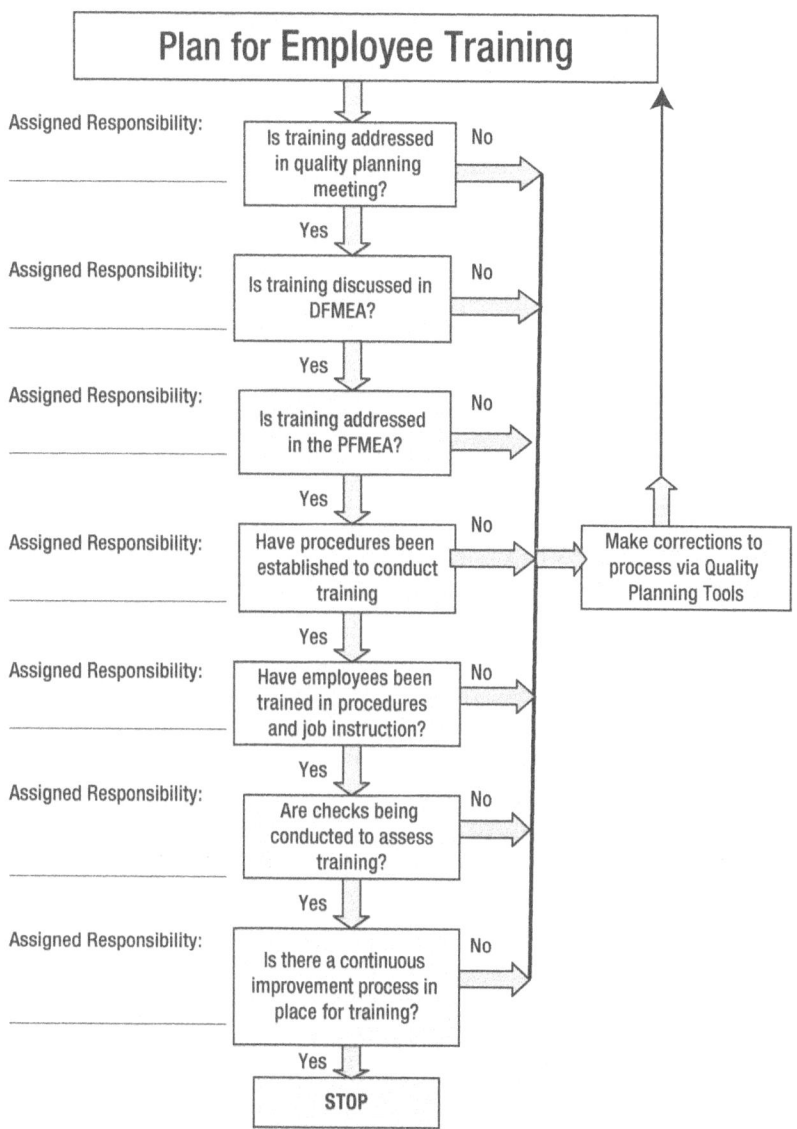

Figure 4-8. Plan for Employee Training

Chapter 4 | Develop a Plan of Attack

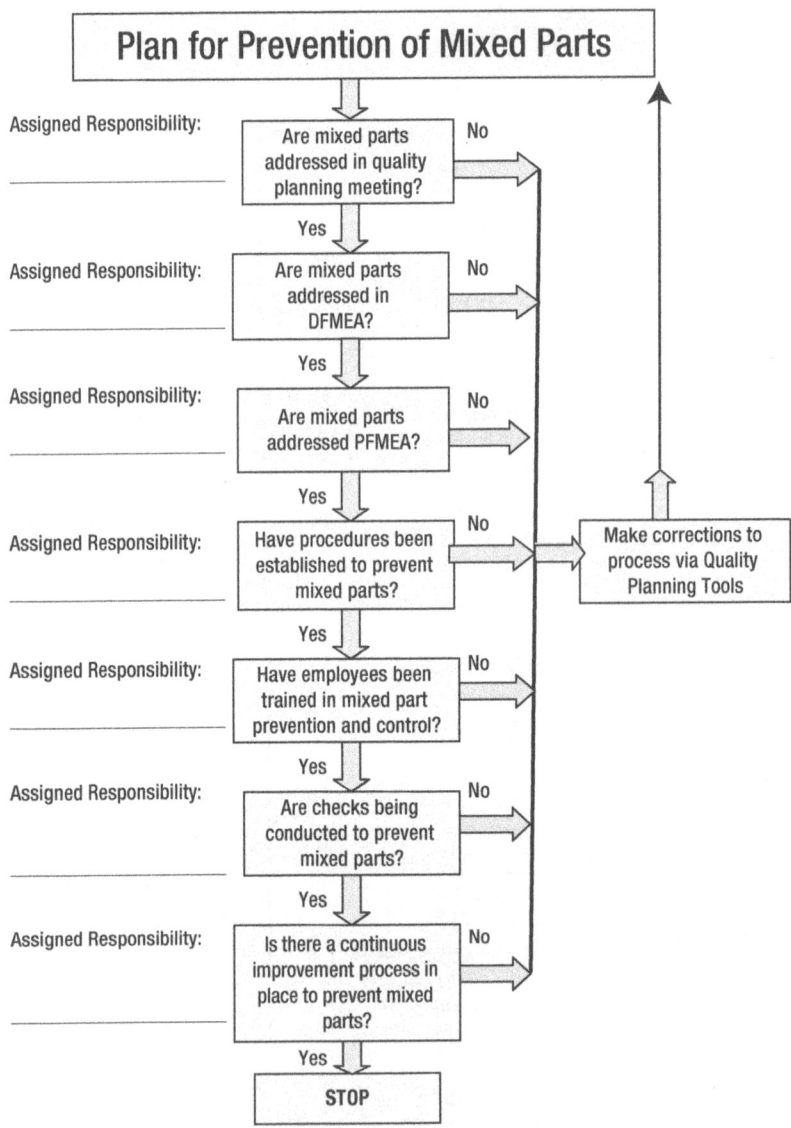

Figure 4-9. Plan for Prevention of Mixed Parts

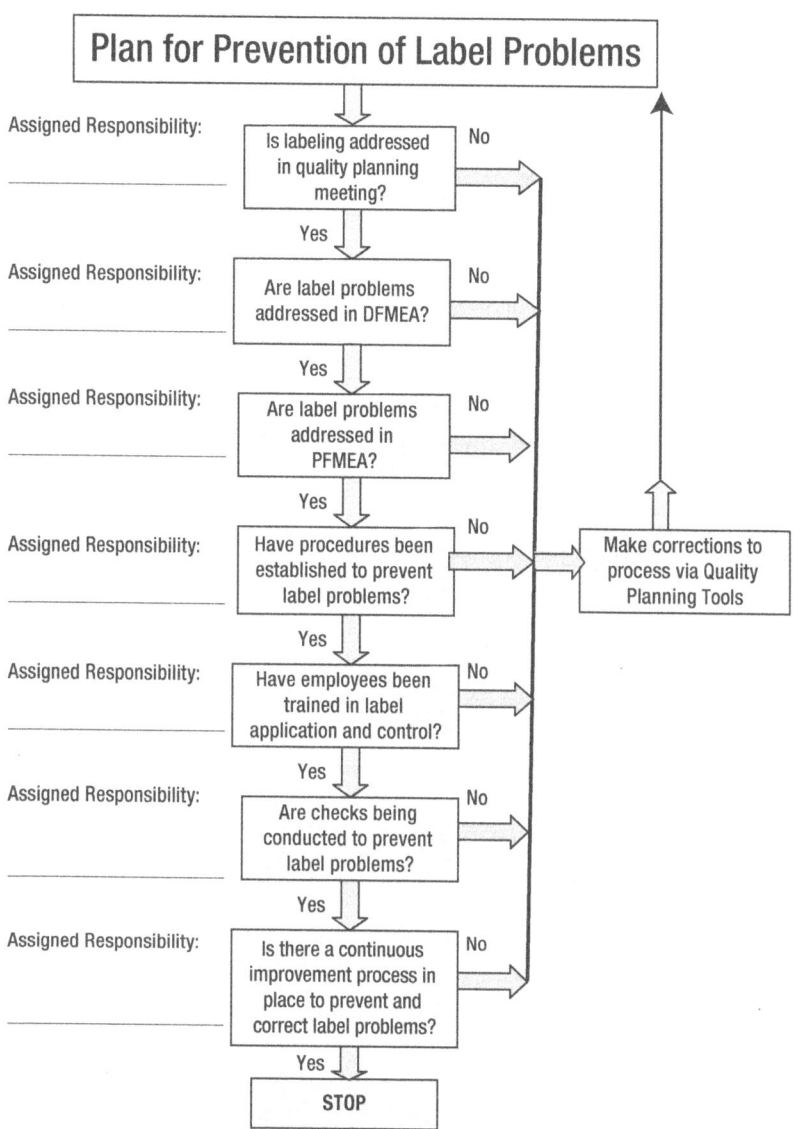

Figure 4-10. Plan for Prevention of Label Problems

Chapter 4 | Develop a Plan of Attack

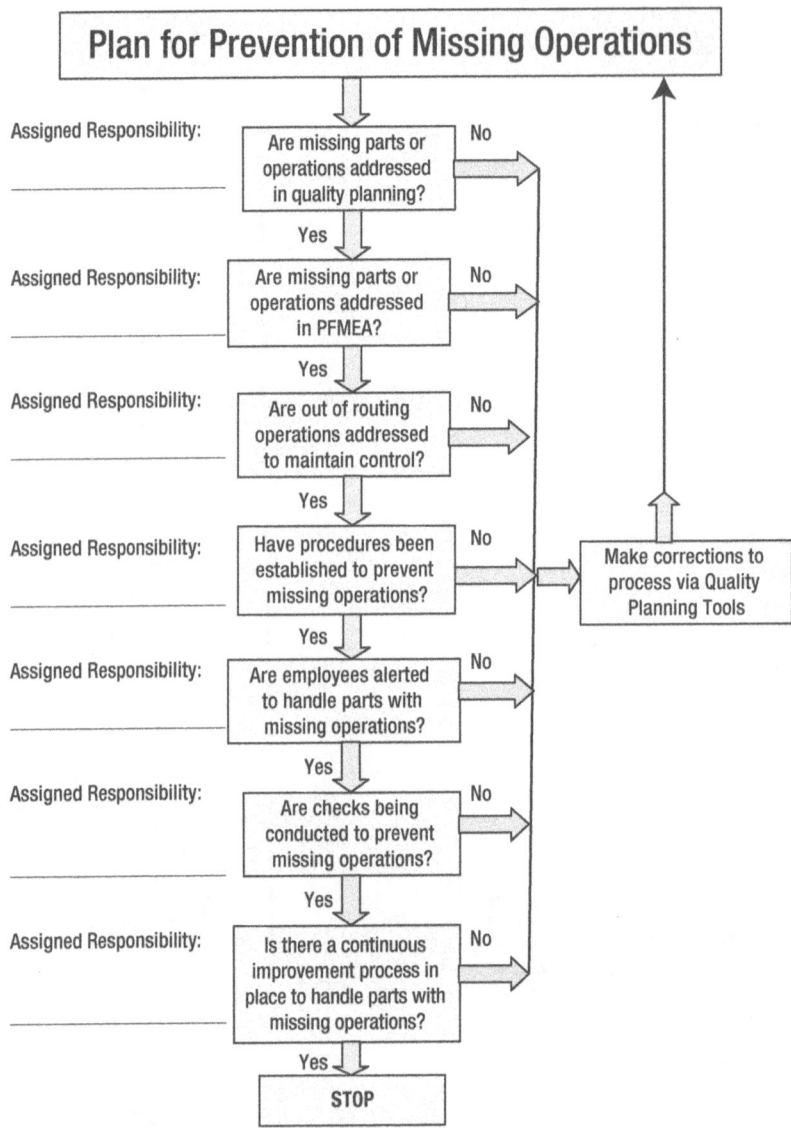

Figure 4-11. Plan for Prevention of Missing Operations

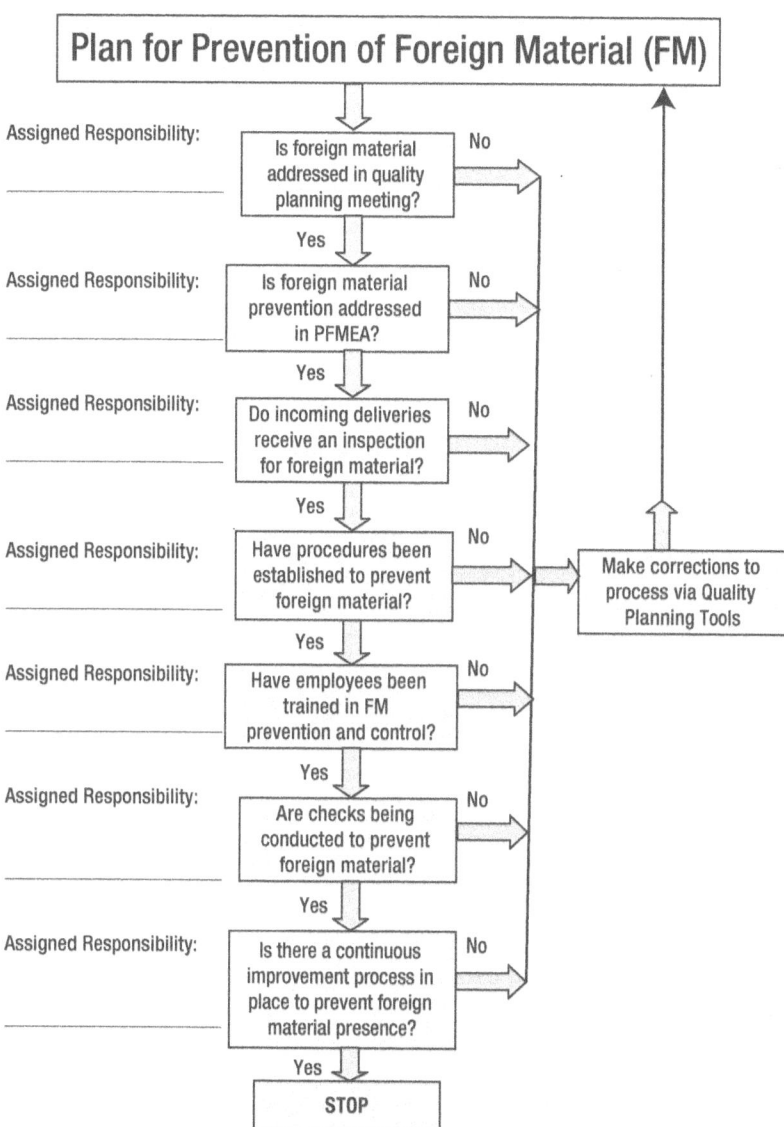

Figure 4-12. Plan for Prevention of Foreign Material

Chapter 4 | Develop a Plan of Attack

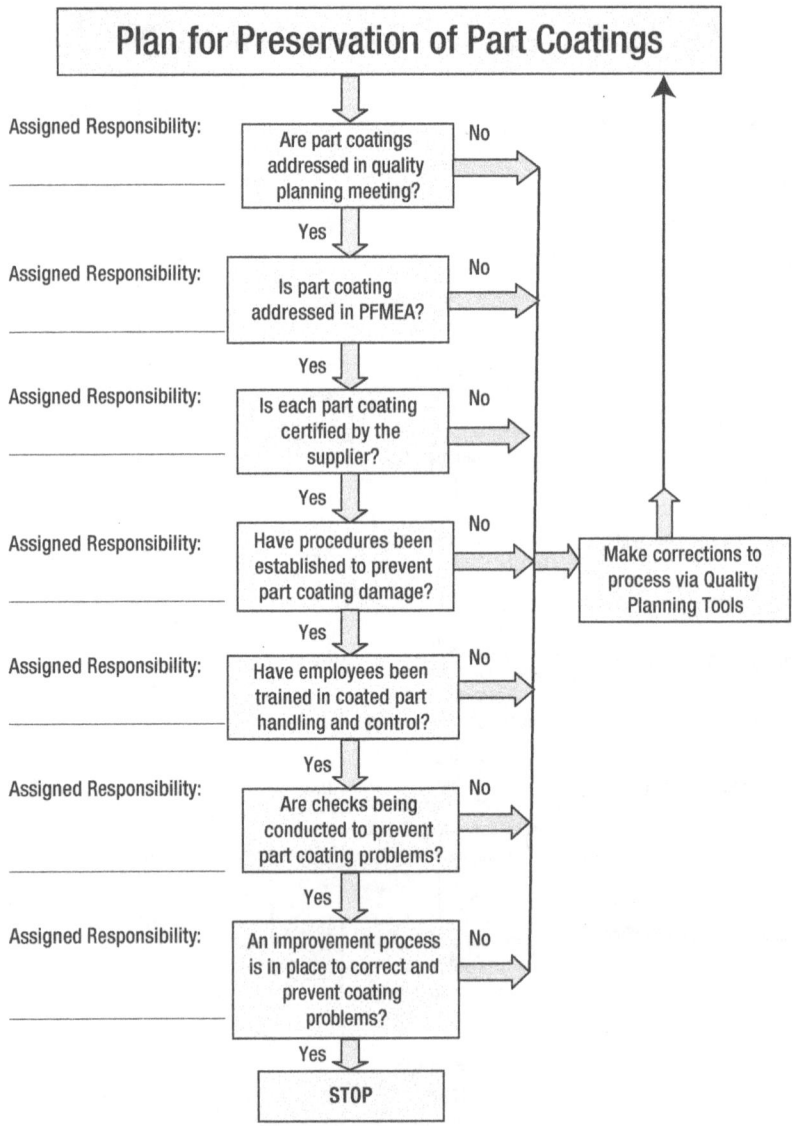

Figure 4-13. Plan for Preservation of Part Coatings

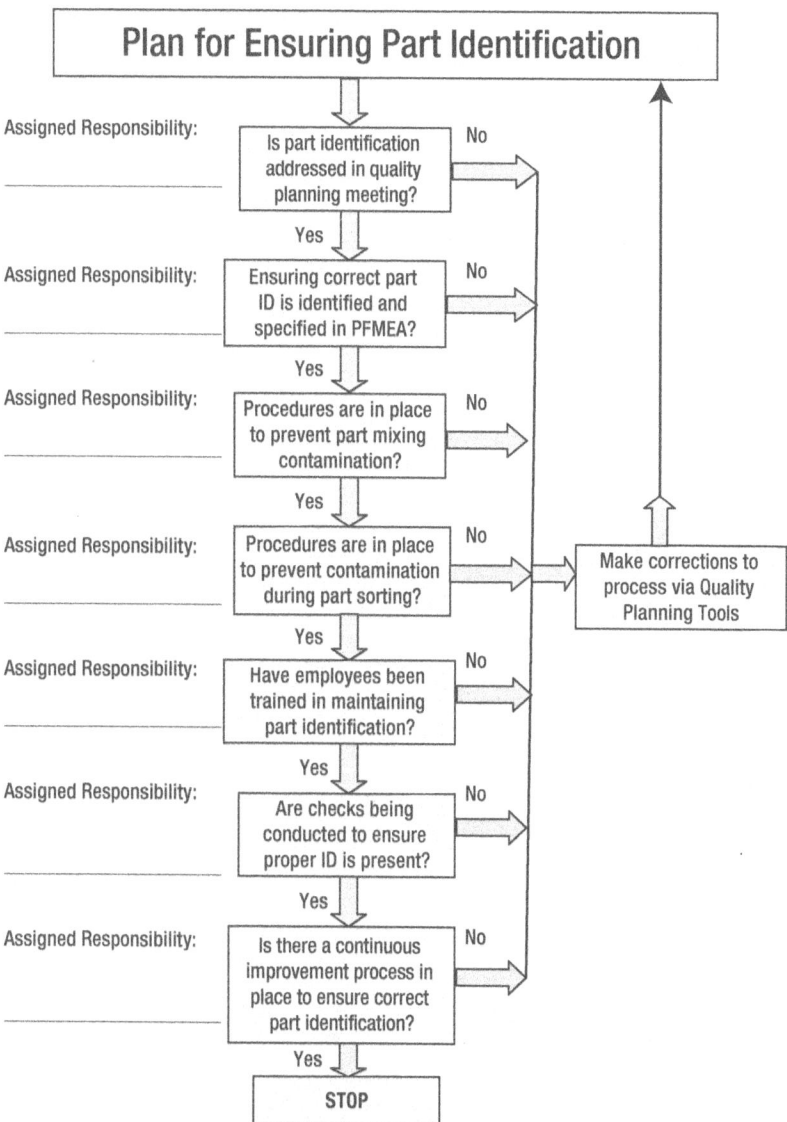

Figure 4-14. Plan for Ensuring Part Identification

Chapter 4 | Develop a Plan of Attack

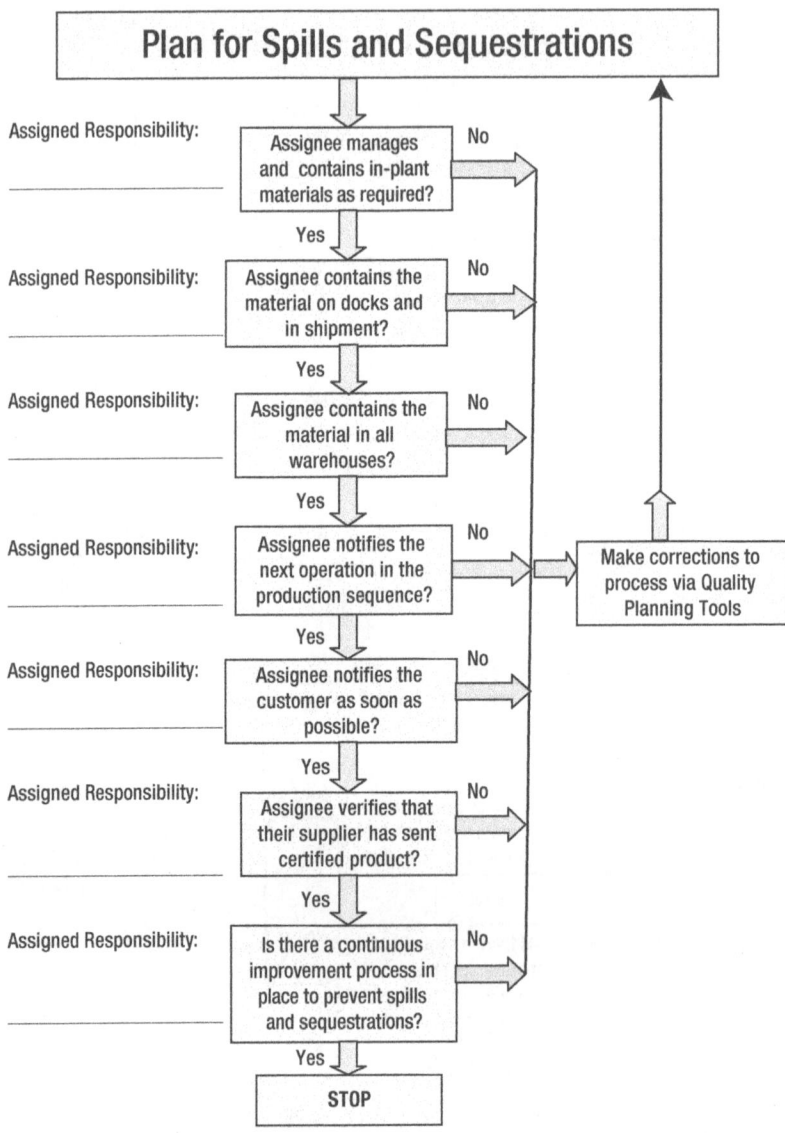

Figure 4-15. Plan for Spills and Sequestrations

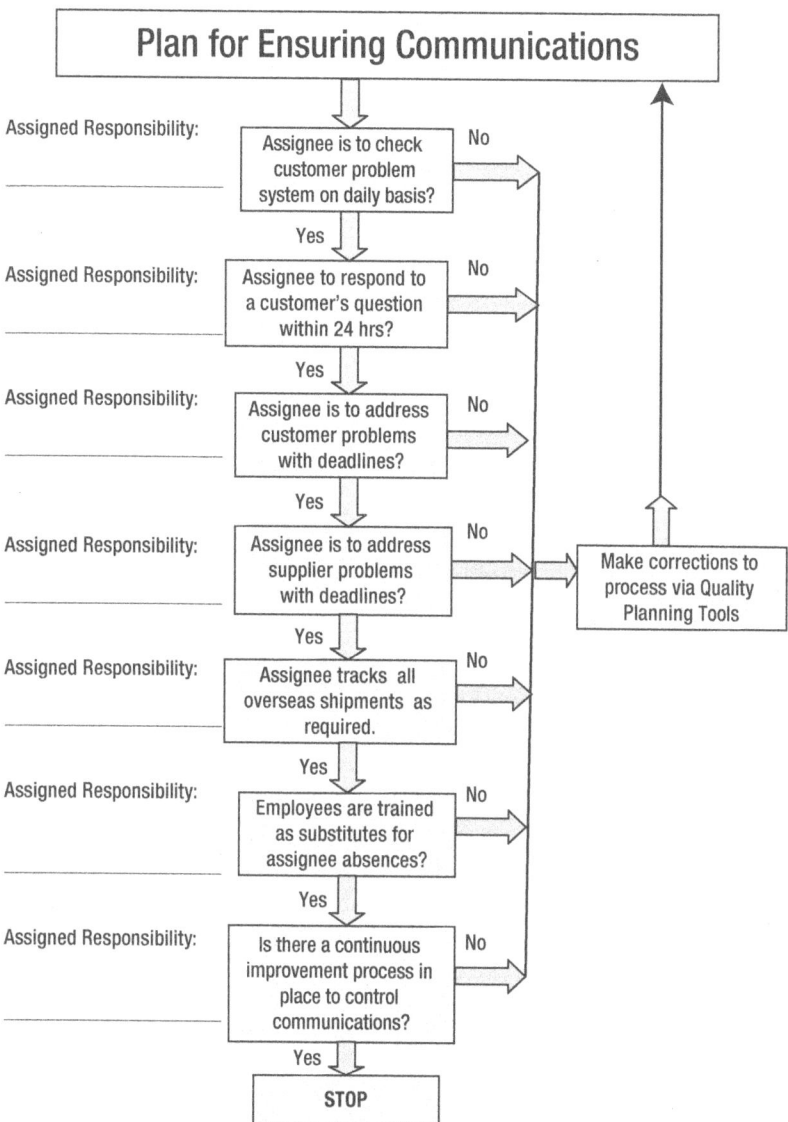

Figure 4-16. Plan for Ensuring Communications

Chapter 4 | Develop a Plan of Attack

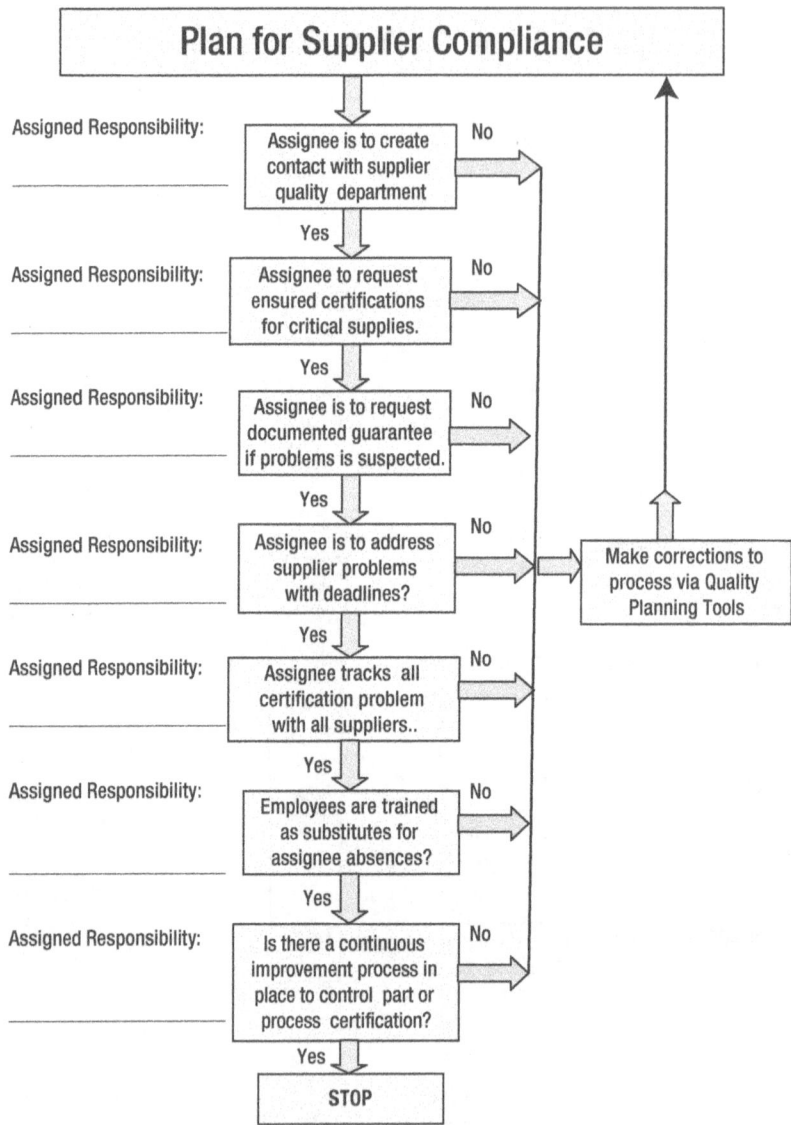

Figure 4-17. Plan for Supplier Compliance

Summary

The individual steps explained thus far can be used as the plan of attack for most problems. In addition, you must take the clues from these methods and determine whether you need to take additional steps in the plan of attack to further analyze the problem conditions. Not to be overlooked is the prevention aspect—this includes assigning responsibility to individuals to ensure responses to detrimental conditions.

The next chapters move to the topic of data collection. You'll determine the means for identifying and collecting data and deciding whether it is significant. We provide examples for generating visual rating systems when measurements cannot be made. As you'll see, you can apply these rating systems to most projects to help determine whether there is a real difference between the good and bad parts.

CHAPTER 5

Collect Relevant Data

Step Five

The fifth step is to collect accurate data that is relevant to the problem you are trying to solve. Data collection is the means of capturing available characteristics for easier analysis. A visual evaluation, which is described in later pages, can also be used to assign values.

The previous chapter discussed the steps for applying a plan of attack and how to determine if additional steps were required to analyze the problem. We looked at the use of simple tools that could generate data from direct measurements or from a visual rating system. We also discussed the importance of assigning responsibility to prevent problems for occurring and providing solutions so that they would not recur. In this chapter, you will further investigate the collection of data and will read about the differences among variable, attribute, and conditional data. You'll also learn how to use a visual rating system.

Past and current data may be required and necessary to solve a manufacturing problem. Such data may help you identify clues that lead to a solution of a problem. It will also aid in identifying which, if any, products should be contained and controlled. All questionable products should be sequestered and certified for compliance. Suspect products should never be passed along to the customers.

Use the Problem Sheet to Gather Data

You can use the problem sheet, shown in Figure 5-1, to collect conditional data when a problem first arises. It requires you to consider the conditions that could be involved in causing the problem. Ideally, you can describe a malfunction as a forming, assembly, or transportation problem, and document whether the part was acted upon with a constant or impact force. That helps to determine whether the part had improper energy applied or if the part was faulty.

Problem Definition Sheet

Describe the condition:_____

_____

List the conditions that are most relevant and observable:

	YES	NO		YES	NO
It is always present?	___	___	Are these virgin parts?	___	___
It doesn't fit?	___	___	Reworked parts?	___	___
It won't function?	___	___	Current shipment?	___	___
Is part too weak?	___	___	All shipments?	___	___
Excessive force was used?	___	___	Only one supplier?	___	___
Handling problem?	___	___	New supplier used?	___	___
Processing problem?	___	___	New material used?	___	___
Assembly problem?	___	___	New design used?	___	___
Has the supplier certified no changes in their materials or processes?				___	___
Can discrepancy be made purposely? (Try to make one.)				___	___
Have two groups of five been saved and compared for differences?				___	___
Have you collected and photographed 5 Good and 5 Bad parts?				___	___
Are there any unusual witness marks or deformations on the samples?				___	___
Does everyone involved agree with the problem description?				___	___

Figure 5-1. Problem Definition Sheet

If an operation failure occurred, I recommend that you replicate the failure on purpose if possible. Ask yourself these questions:

- Was there a kinetic or potential force applied?
- Was the problem caused because the force was too great or because the part was too weak?
- Is the failure repeatable?
- Is the point of origination clear?
- Was a comparison made of good and bad parts?

Very important clues are provided if you can describe the failure; it leads to understanding the failure mode.

The questions in Figure 5-1 require simple answers that will help you capture, control, and identify the important data related to the suspect parts. The conditional differences may lead to your identifying that the assignable causes lie not only with the failed part, but also with the materials, machines, tests, methods, procedures, or environment.

Collecting these types of clues provides viable data that will expedite a solution.

After you achieve consensus on the fault being studied and identify the peripheral information, you need to evaluate the defect. Enumeration using acquired variable (number) data is more precise than comparing collected attribute (description) data, because the information is more defined and provides greater measurement certainty. For example, an inch, a gram, a mile, and a hectare can be duplicated around the world with a verbal description because they can all be related to a universal standard of measurement as determined by the U. S. Bureau of Standards. Each of these is variable (numeric) data that requires specific properties. On the other hand, if a customer calls India, China, Mexico, or someone in the U.S. and asks for a deeper red than the sample they received for approval, there will be a different red provided from each location. This is because the attribute data is determined by the visual interpretation of the perceiver and is not specifically defined.

If the assembly calls for a flush mating surface to within 0.001 inch, you know that gauge measurements that exceed 0.001 inches are not acceptable. In this case, the gauge might be a micrometer that was used for the measurement. This is variable or number data. On the other hand, if the assembly requires that the object be painted a robin egg blue at final assembly, you must compare it to and match a previously accepted standardized visual sample of plastic or paint chip to be acceptable. This is attribute or descriptive data.

Once you've defined the defect, you need to determine the meaningful measurements that are related to the problem. Since accurate measuring systems are a prerequisite for data collection, you must carefully choose

Chapter 5 | Collect Relevant Data

the measuring system that's the most applicable. Both variable and attribute systems can generate usable data, but only the variable data can be used for statistical calculations. The attribute data only describes the fault's features.

Collect Variable Data

Variable data collection devices should be subject to gauge reliability and repeatability studies before use. A gauge that exceeds 30% gauge error as determined by industry-accepted methods is unsatisfactory. Therefore, the measuring instrument must be calibrated to make the measurements over the measuring scale required, without biasing results.

Sensitive measuring instruments can be damaged by sudden impact or even temperature differentials. A micrometer that is dropped to the floor may be severely damaged and cause inaccurate readings if not properly recalibrated with an established standard before reuse. Calibration at timely intervals is a precautionary measure to help ensure that measuring instruments and gauges are up to standard and acceptable for use. It should be noted that there are other conditions that can affect the accuracy of measurements, including vibrations, direct sunlight, drafts, contamination, and other environmental conditions. However, anyone making measurements should regularly calibrate to ensure gauge accuracy.

The direct effects of the variables are sometimes easily determined by observation. You should inspect any defective parts vigorously for every clue. Not only should the part be inspected for discrepancy characteristics and location, you need to capture, record and compare any pertinent information. This might include, among other identifiers, the following:

- Serial number
- Pattern number
- Date of manufacture
- Lot number
- Surface appearance
- Finish
- Discoloration[1]

---

[1] Note that a *variable* is something that is apt to change. Pattern serials, dates, surface finish, and lot numbers, for example, vary—as do the other characteristics mentioned. Surface appearance and discoloration are attributes, but they are also variables that cannot be quantified easily with variable measuring instruments.

In addition, be cautious. Repeat measurements might not be the same due to fatigue in a prior test. For example, an air-filled bladder may measure as larger or longer than it does when it's deflated. To eliminate any potential causes, you must first compare the acceptable and unacceptable conditions for observable differences.

Collecting Attribute Data

Attribute data is used to describe characteristics that cannot be collected from variable measurement devices. These characteristics include color, taste, brightness, discoloration, rust, and so on. Attribute data is generally collected to evaluate a condition that doesn't lend itself to variable instrument measurement. Attribute data involves estimation and involves visual, auditory, or other sensory perceptions.

You can make your attribute data more useful by creating an evaluation system for comparisons. This might be in the form of assigning a designated comparison value and rating system.

Note Attribute data does not provide as accurate a description as measured data, but it is helpful in measuring comparisons that cannot be adequately measured with a measurement gauge or device. The amount of applied grease, the proliferation of holes or porosity in metal castings or glass, the gradients of color—these are all examples of important attribute data.

Visual Evaluation System

Figures 5-2 through 5-4 are examples of visual evaluation systems. They have been used to contrast the amount of scoring on gear teeth, determine the initiation of gear tooth pitting, record leakage rates, compare amounts of grease present, match differences in defect bleeding, rate presence of voids, compare ignition discoloration, establish porosity levels in castings, and so forth. You can create one, on an as-needed basis, for any class of problem you might have. Such systems allow you to generate clues when numeric measuring systems are not available or tenable.

Chapter 5 | Collect Relevant Data

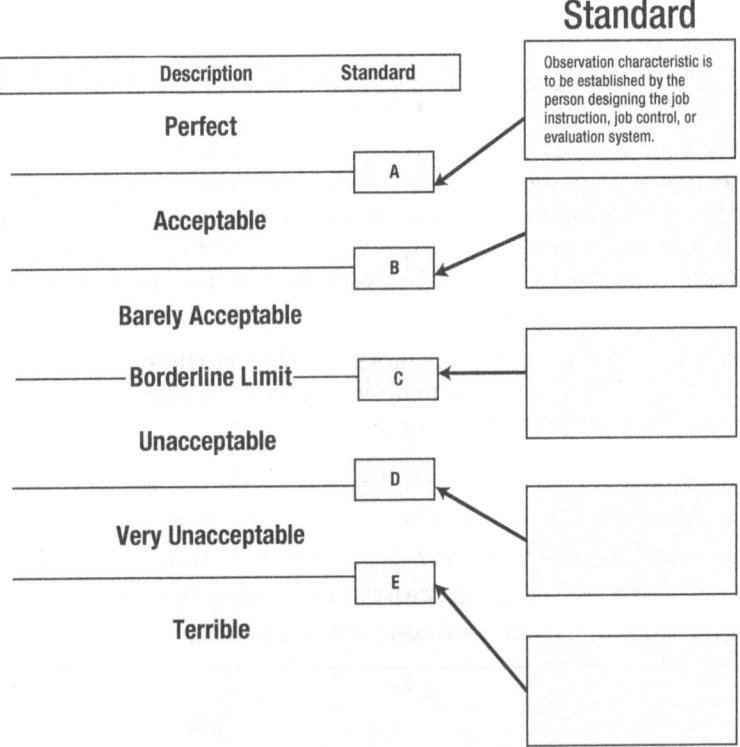

Figure 5-2. Visual Evaluation Form

Visual evaluation systems are easily constructed. You can describe the entire range of a characteristic using six classifications. Each of the classifications has a specific acceptance rating, ranging from perfect to horrendous. In addition, there are five standard ratings—A thru E—that are determined by the person designing the system. They indicate the conditions that may be present. Standard A depicts an almost perfect part, whereas standard E may represent the most unacceptable condition found. The rating of C should always be the breakpoint between what is acceptable and what is not.

You can also use a visual rating sheet to enhance operator instructions, whereby the actions of the employee are guided by a visual rating sheet posting. This is shown later in the chapter as a means to control grease application. By observing the posting, the employee knows whether to continue the operation, adjust the system, or notify a supervisor who has the authority to react to the system.

It doesn't require undue expertise to develop these standards. You simply collect a number of the parts over time related to the characteristic under study and attempt to capture examples of the best and worst conditions. Choose the sample that is barely acceptable, and call it Standard C. It differentiates between good and bad parts. Then choose two samples on either side of Standard C; they differentiate between a perfect sample and the worst sample. Arrange these four samples in order of severity and rank them as A, B, C, D, or E, according to your rating system.

Once you construct the system, collect current samples and rate them with a team or individually. If they are less acceptable than the standard B but more acceptable than the Standard C, you could rank them as C+, which would be an acceptable part. If they are more acceptable than Standard D but less acceptable than Standard C, you could rank them a C−, which is an unacceptable part. This artificial comparison system will allow you to make decisions that will aid in the problem solution.

The next pages show a visual evaluation system that is used to evaluate porosity in an engine front cover (see Figure 5-3) and one that defines the rating system for acceptable grease coverage on incoming seals (see Figure 5-4), which affects engine leaks.

Chapter 5 | Collect Relevant Data

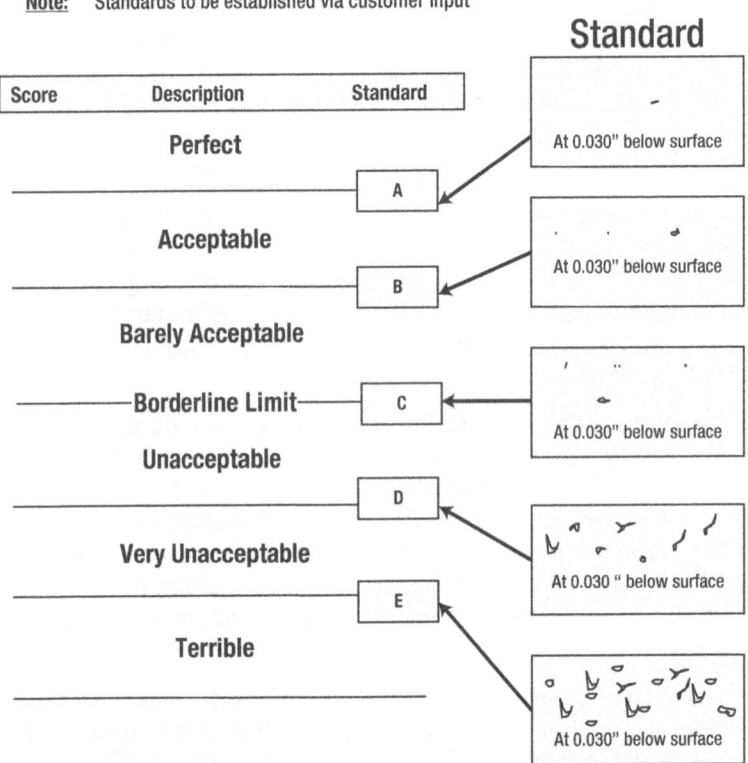

Visual ranking systems can also be set up to establish job instructions and process controls. They can be posted on job sites to prevent manufacturing problems.

Figure 5-3. Example of Visual Rating for Porosity

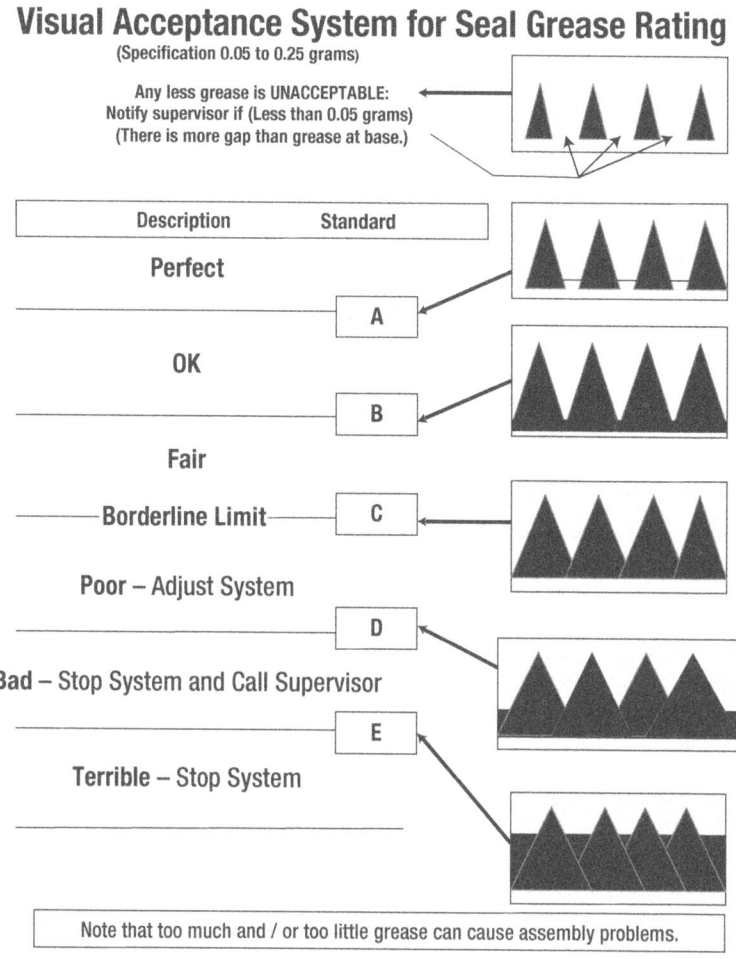

Figure 5-4. Example of Visual Acceptance System for Grease

Under ordinary circumstances, a component may not be inspected for porosity. The amount of porosity shown in these standards exceeds the allowable amount in engine assembly components, but would be appropriate for plastic horseshoes used in a children's game set. It depends upon the circumstance and product use. In the case of the horseshoes, excess porosity resulted in a broken horseshoe when it was tossed to a retaining stake. The porosity, which is basically the presence of voids in a product, can induce leaks, failure planes, or structural weaknesses.

Figure 5-4 shows the sample that was referenced earlier. It provides a visual instruction to the system operator about the appropriate level of grease. The operator can continue operating the system, adjust it, or call the supervisor to terminate it. In this case, the supervisor's actions are not explained; she

Chapter 5 | Collect Relevant Data

might have installed some additional action on the line after adjusting the system, created a rework station to salvage the non-conforming product, or terminated the system. The initial instruction, though, was to help the operator make informed decisions.

Summary

In this chapter, we further investigated the collection of data and described the difference between variable, attribute, and conditional data. Variable data is obtained by direct measurement of a component's characteristic, feature, or dimension. Attribute date is generated through the measurement system using visual ratings, and conditional data depends upon the conditions encountered. The visual rating system can be used to generate data or as a means of controlling the quality of the product or process; the data it can generate depends upon the evaluator's aims and perspective.

In the next chapter, we will cover the six main actions necessary for success. We've identified these six actions during our work in mentoring 200 suppliers to identify, correct, and prevent manufacturing problems.

CHAPTER 6

Generate Clues

An Interlude

The previous chapters addressed five of the eight steps in the problem-solving process. These and the remaining steps give you a macro view of industrial problem solving.

This chapter further explains those thoughts but addresses the subject in a micro setting. This is necessary, because some problems require more than a simple comparison to solve them. Some problems require an in-depth look. For example, a visual review of a part might not be enough to determine the problem unless you observe it with a magnifying glass or a microscope. This is moving from a macro view to a micro view. In a similar situation, it might be necessary to measure and define specific dimensions or even to use unconventional methods to identify, correct, and prevent manufacturing problems.

Hopefully, the text and examples in this chapter will provide you with insights. Clues, as you will see, can be generated in many different ways. There are many types of charts that you can use to elicit valuable clues, for example. The Pareto chart, the run chart, and the concentration diagram are three common but important tools. These lie outside the scope of this book, however, and you can find them well described elsewhere.

Generating clues for problem-solving success hinges upon the six actions discussed earlier in the book. I have employed these actions over and over again in solving quality problems while working with over 200 suppliers of an automotive engine manufacturer. They provide micro information in the form of text and additional tools. These six action items are:

1. Define the specific problem and condition (the characteristics or physical appearance of the subject) thoroughly. Evaluate problems by the dollar loss that they generate.

2. Verify that the process is operating as intended and specified. Conduct a review of the control plan and the operating procedures to ensure that the process is being managed as desired.

3. Observe the operation for detrimental or ruinous conditions. Determine if there is anything specific causing distress in the process.

4. Develop a check system/assessment plan for recognized manufacturing control items. Decide what is important to control so that the process is acceptable.

5. Conduct checks with more than one layer of inspection.

6. React to the conditions that must be corrected or improved. Add corrective actions to the audit list if any item or condition is found that has or will have a negative effect on the process or the product.

Define the Problem

Defining the problem is the most critical step, and the methods of discovery are explained in Appendixes A, B, C, D, and E. The initial steps to be followed were presented in the problem corrective action worksheet, shown in Chapter 1 and the problem definition sheet shown in Chapter 2.

Sometimes it is necessary to review and reinforce important concepts, even when they've been previously discussed. Let's make certain that you have a good understanding of the terms "problem definition" and "problem cause" before we move on.

An illustration with examples is probably the best way to overcome any misunderstanding. Three examples come ideally to mind. The three problem definitions are as follows:

- The floor after mopping is not clean.
- The machined surface is not smooth.
- The crankshafts are warped.

Note that each of these indicates an unacceptable condition. The problem definition does not attempt to specify the cause of the problem because the cause is unknown. And it generally cannot be known at this point.

Allow me the liberty of specifying the causes of the three conditions to illustrate the examples and to simplify the definitions. The cause of the first problem may have been the use of an oil-contaminated mop to clean the floor. The cause of the surface not being smooth on the machined part may have been due to incorrect speeds and feeds on the machine. The cause of the warped crankshafts may have been due to an improper foam coating and drying operation that resulted in a warped, lost foam pattern that allowed the iron to take the same warped shape.

Defining the problem is an initial step that should have the most critical recognized adverse effect highlighted. Attempt to make this as micro as possible. The floor was dirty after mopping. Or simply, the floor was still dirty. This focus defines the problem in a micro format.

Charts and data are useful tools in acquiring clues to solutions. Some measurable and observable differences that can yield clues are variations in:

- Seal OD (outside diameter) measurement
- Fuel pipe air test leak measurements
- Tubing circularity measurements
- Head exhaust chamber leaks
- Machining energy applied
- Concentricity of crankshaft main bearings
- Part fits/configurations
- Part finishes
- Other process variables and variations
- Porosity or void presence
- Crack location and severity
- Almost any other characteristic imaginable

Note Charts can consist of simple graphs, sketches, or depictions that describe information available from a visual investigation. For example, a graph that shows two lines of the same product, one of which has twice the number of rejects, is a simple but telling display. A chart can simply be displaying the difference in percentage or frequency of something. Don't limit yourself, therefore, to preconceptions of what constitutes a "chart."

Verify the Process

Verify the process to ensure that it is operating as intended. Are all of the process specifications in accord? Observe and record—if they're available and applicable—speeds, feeds, transfers, temperatures, chemistries, heat treat, static controls, internal materials, supplier certified materials, and so on. You must check anything that has been established to initiate and approve the original production run and updated improvements.

Observe the process for detrimental or ruinous conditions. They should not be observable while walking from the start to the end of the process. If they are, you have some clues that will help you solve the problem. Sometimes these conditions are what cause component breakage or damage at transfer or separation points. They can include uneven oven temperatures, open window drafts, roof leaks, jams at transfer points, power interruptions, double machine strikes, and so on.

What does a detrimental condition look like? Say a damper manufacturer provided a product that was displayed under the hood of customized vehicles at car shows. The car owners desired a product with a perfect appearance that did not have blemishes. One day, scuff marks suddenly started appearing on some of the dampers. Observing the assembly line process allowed us to recognize the cause of the detrimental condition—units that remained on the line during lunchtime with the power rollers running were scuffed up.

The power rollers were coated to prevent scuffing but the trapping of the damper at a transfer point directly above the power rollers allowed the scuffing to occur. Again, the problem definition was that the "dampers were marred." The cause of the problem was that the parts were trapped over the power rollers.

Develop an Assessment Plan

Develop an assessment plan to capture variables that should be controlled in the process. We will discuss check plans[1] in later chapters to show how they should be designed and conducted. All important items required to certify the process should be included from the control plan,[2] the DFMEA, and the PFMEA. This includes specific tools, methods, paths, operations, procedures, and so on.

---

[1] A *check plan* is an inspection of specific process variables to ensure compliance with the methods used to conduct an operation. It can be a simple audit for detrimental conditions, or a visual check to determine if the specified and authorized work is being performed. It is similar to a safety check that verifies that all employees are wearing safety glasses in designated areas. However, a check plan can contain any number of observations to be conducted by the checker.

[2] Again, a *control plan* is a set of operating instructions that the manufacturing or service operation must follow. It can specify the materials, methods, checks, inspections, identifications, tools, or environments required to perform the production or the service. Its purpose is to ensure understanding and control of the process under scrutiny.

Industrial Problem Solving Simplified

An assessment plan should include provisions that protect against allowing detrimental conditions to recur after they are once recognized. In the example of the scuffed dampers, an audit of the production line might include an assessment plan with instructions to shut down the powered rollers at line stops such as breaks and lunch. They might also audit to ensure that there were no trapping areas where the dampers would be contained above a set of moving power rollers. Someone conducting the assessment plan would verify that these system precautions and controls were satisfactory to prevent damper damage.

Conduct Checks

Conduct assessment checks with more than one person responsible for the accuracy. This can include an assembler, quality representative, supervisor, or manager. Review results help you control the established procedures and required steps. Multiple or staged checks—two or more—provide verifiable results that are representative of the conditions. (The rationale for this requirement is explained in a later section.)

The assessment checks survey the operation for situations that allow detrimental conditions to occur. They also verify that the established controls and operating procedures are being practiced. The reason it's best to have more than one person perform these tests is to avoid tester bias. If one observer has a blind spot to a particular condition, you still have the other tester who will likely recognize the problem, so that it can be addressed and corrected.

React to the Problem

React to detrimental conditions as soon as they are identified. Whether it requires revisions in job instructions, tool use, or better inspection processes, it is important that once a problem is identified, a corrective action be developed and enacted for all future audits so that the condition doesn't recur.

Depending on the type of fault encountered, the methods outlined here, along with these six steps, will help you develop a list of clues to solve most problems. Additional steps are provided in an example, which will help to focus the clues and provide a solution.

Example Manufacture Design

Suppose that you are going to manufacture yo-yos, and you want to manufacture a toy that will work for the greatest number of customers who want to perform complex tricks.

Chapter 6 | Generate Clues

You must first consider the functions that you expect the yo-yo to perform. It should be able to perform tricks like "sleeping," "around the world," "walk the dog," and "baby in a cradle."

The "sleeping" trick happens when the yo-yo's central shaft, which connects the two halves, spins for more than three seconds on a fully extended string at the floor level. It does not come back to waist level until the hand controlling the yo-yo offers a slight nudge.

"Around the world" requires that the yo-yo enter a sleeping state as well, but it differs in that the yo-yo is not tossed downward to achieve the sleeping state. It is tossed from the waist area to a position above the head so that the momentum of the toss results in the yo-yo traveling a full 360-degree rotation before it is given a nudge to return to the tossing hand (see Figure 6-1).

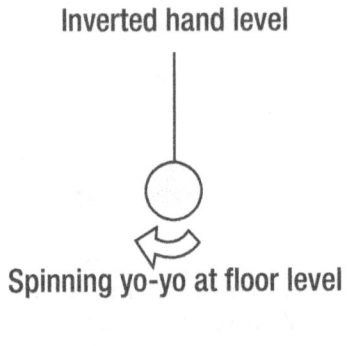

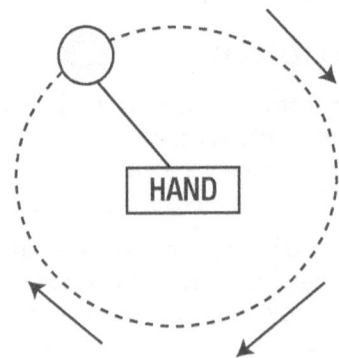

Figure 6-1. Yo-Yo Trajectory for the Around the World Trick

(The other tricks are similar in nature but involve steps that are not critical to its operation.)

You must now consider the requirements necessary to allow these conditions to exist. In order to gain experience in determining characteristics that are important to the manufacture or operation of a device, you'll now attempt to develop the requirements for a yo-yo. Now is a good time to attempt this. Before going further, list the items and attributes that you consider necessary to allow the yo-yo to operate satisfactorily.

Note This type of training is useful whenever you approach any manufacturing problem in the future. It will sensitize you to some of the important considerations for any problem. The more times that you attempt to list those characteristics that are involved in a problem, the more intuitive your attempts will become. Each experience should provide new considerations that would not be possible if the attempts to make the definitions had not been tried.

Here is my list. How does it compare to yours?

- The yo-yo must have adequate strength to withstand the toss.
- It must have adequate mass (weight) to be tossed.
- It should have adequate mass to achieve the angular velocity required to sleep.
- The central spool piece must be machined smooth.
- The material must be balanced.
- Both sides must be chamfered to prevent string wear.
- The string must have a low coefficient of friction.
- The string must be strong enough to withstand the toss.
- The string must have a smooth surface.
- The string must not be too short or too long.
- The string must not be permanently attached to the shaft.
- The shaft (spool piece) must be inserted within the loop of the string to allow rotation when the unit is sleeping.
- The operator should be a certain age.
- The manufacturer should provide adequate instructions.
- The user requires training.
- Assembly must not contain sharp edges for safety.

This list is not complete. If you listed five or more considerations, you are on your way to effective problem solving. Regardless of the list you developed, be assured that your experience will increase every time you apply this technique to understanding a problem.

In reality, solving a manufacturing problem requires less analysis than what is shown here. In most cases, it's easier to determine the cause of a flaw than it is to list all the design requirements. You observed this type of simplified analysis in the poor fusion weld concept sheet example in Chapter 3, as well as other examples on the previous pages.

Once you have identified the conditions that can potentially cause a problem, and connected them to the problem being investigated, it's easier to recognize the conditions and address them.

Guidelines for Clue Collection

Before you start any hard data collection, you must verify that all existing controls and work procedures are in place and being followed properly. Altered work procedures can result in defects. This important step is often overlooked in the rush to solve problems. The problem definition sheet discussed in Chapter 5 provides you with a quick, cursory check of internal operations. Use it before you look in other areas.

Next, be sure all parties agree on the problem definition. Then, observation can provide clues regarding the direct effects of variables, which is sometimes all you need.

Suppose the customer desires a smooth finish on the machined product that he purchases and is dissatisfied with the current finish. The problem may be stated, "The product is not smooth."

Unless everyone understands the problem description, you may find that some people want to focus on the fact that the part is discolored or should have a label. In some cases, it's the lack of knowledge of a product, and in other cases, it's the lack of knowledge of the customer's requirements. Regardless, this miscommunication impedes progress.

Defective parts should be inspected vigorously for every clue, and not only for their characteristics and locations. You should record all pertinent identifiers, including serial number, pattern number, date of manufacture, machine path, lot number, surface appearance, finish, and any discoloration.

Once you have taken these steps, you can narrow the investigation's range. Save only the essential variables listed in Figure 6-2. Cross out the variables that are not applicable and then scrutinize the ones that remain.

Problem Definition Sheet

Describe the condition:_____

_____

List the conditions that are most relevant and observable:

	YES	NO		YES	NO
It is always present?	___	___	Are these virgin parts?	___	___
It doesn't fit?	___	___	Reworked parts?	___	___
It won't function?	___	___	Current shipment?	___	___
Is part too weak?	___	___	All shipments?	___	___
Excessive force was used?	___	___	Only one supplier?	___	___
Handling problem?	___	___	New supplier used?	___	___
Processing problem?	___	___	New material used?	___	___
Assembly problem?	___	___	New design used?	___	___
Has the supplier certified no changes in their materials or processes?				___	___
Can a discrepancy be made purposely? (Try to make one.)				___	___
Have two groups of five been saved and compared for differences?				___	___
Have you collected and photographed five good and five bad parts?				___	___
Are there any unusual witness marks or deformations on the samples?				___	___
Does everyone involved agree with the problem description?				___	___

Figure 6-2. Condensed Problem Definition Sheet

The evaluator should ask and answer the following questions:

Is the fault always present at a continual level, or does a condition spike as a periodic happening? Is the flaw related to the part shape (geometry) or was some type of power present (force)? Did something go snap (incident), did it contain a flaw (fault), or is one of its attributes (trait) suspect? Is the part (strength) too weak or was excessive (force) power applied? Was any work applied or was it a static failure? Is the failure more related to machining, assembly, or design? Is the condition a cause of scrap or rework? Is visual breakage present or was it a malfunction? Was the defect the result of handling or processing? Did the assembly or a component fail? Are witness marks present, and where are they located? Does this happen on all shifts and machines the same? Has the supplier swore (certified) that the process or materials have not changed? Has supplier certified that they and their supplier(s) have made no unauthorized changes also?

Each of these considerations is key in pinpointing conditions that result in defects. You can expect your supplier(s) to provide you with pertinent information if you encounter problems that are beyond your observation.

Other things to consider: has the design changed? Is this a mechanical or supply problem? Are these virgin or reworked parts? Each of these divisions provides information and narrows the investigation required. The answer to each comparison focuses attention on a narrower view. Any remaining characteristic could provide insight to the source of the unfavorable condition.

Identify and Sequester Bad Parts

More often than you think, scrap or unacceptable parts are placed off to the side without being properly identified. These parts invariably find themselves back in the processing system and may eventually be delivered to the customer. This occurs because:

- The parts were not recognized as being noncompliant.
- The parts were placed back into a container while still noncompliant.
- Only one of many nonconformances present was corrected.
- Proper work instructions were not followed.
- Checks were not conducted to ensure procedure compliance.

It is best to immediately tag, paint dot, or damage a nonconforming part in accordance with the work instructions. Hopefully, this will prevent a nonconforming part from getting to the customer. If it does end up back in the processing system, marking it will help to trace and identify it.

Moreover, you should permanently identify reworked parts for future recognition. Be sure to specify when the certified work was done. You must conduct checks to ensure that people comply with established procedures for routine and rework operations. These instructions should include adequate checks to ensure that the intent of the operation is fulfilled and that the results meet the specifications. All parts must be adequately controlled.

Parts that are lying on the floor should be considered nonconforming. They should be sequestered, identified, and sorted. Simple fasteners placed in a wrong container can cause automatic feeder jams and cause downtime at customer locations.

Sometimes, you can apply a small paint dot to a part requiring future identification. Be sure to apply a dot to an area that's not readily visible to avoid detracting from customer satisfaction. The dot can also be used in problem-solving applications if it contains a fault. Not only can it be used by the supplier to ensure compliance, it can also be used to perform problem analysis. Paint dots can be used to indicate the following:

- A rework operation has been completed
- A leak check has been performed
- A special operation has been performed
- A previous operation component is present at the next station
- All components are confirmed to be present before final packaging
- A specific item is in place and is to size

If possible, the evaluator should collect five best and five worst examples of the problem. A comparison study is described later. (An explanation of the rationale used is in Appendix E.) For now, you can use these 10 samples as a basis for visual comparison. Describe the dissimilarities of the five best and the five worst. Do all the worst ones have a witness mark? Are all the best ones a certain color? Do all parts measure the same? What are the differences?

Differences between the best and the worst may be extremely small. Nevertheless, if all the acceptable parts don't have marks and all the unacceptable parts do, these marks might be a clue to the problem. Always try to obtain samples from opposite ends of the quality distribution spectrum, as this will make the differences more apparent.

Next, if possible, attempt to replicate the failure. In so doing, you may be able to verify insufficient part strength or the use of excessive force. The technique of identifying the differences between acceptable and nonacceptable parts, processes, or actions further reduce the time you need to provide a solution. In addition, you can use paint dots if there is more than one flow path. The process must be structured so that parts that travel through different processing equipment are identified for later analysis. This is true when the operation is performed at different stations, different machines, on different gauges, or on different lines. The individual flow path or process step identification is a major clue generator.

Use a Defect Diagram

A defect diagram can be a considerable source of information and may illustrate where a fault is most prevalent. Figure 6-3 was used to evaluate catastrophic crankshaft failures in an auto assembly plant.

Chapter 6 | Generate Clues

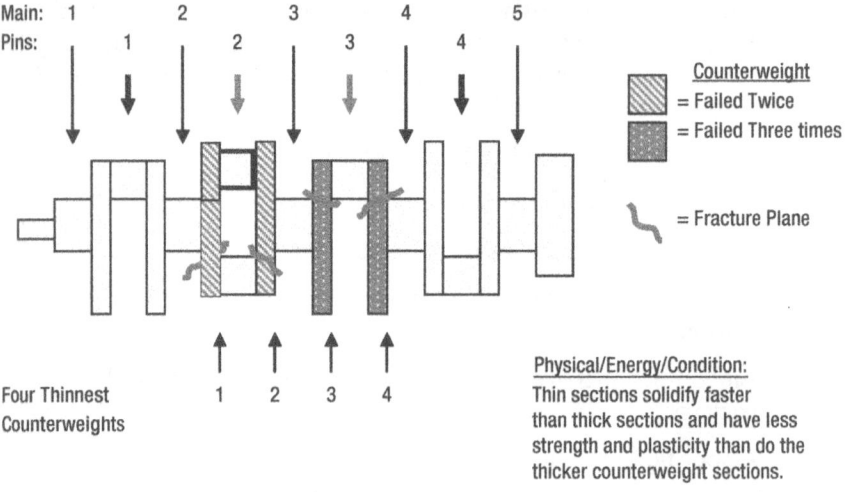

Physical/Energy/Condition: Thin sections solidify faster than thick sections and have less strength and plasticity than do the thicker counterweight sections.

Failure: Fracture occurs adjacent to pin diameter at the interface with the main bearing where major sectional cross sectional changes occur. The undercut in the crank may be start of break.

Crankshafts dropped in the cherry red state were present at a rate of 8 per hour. Some of the crank failures contained "hot cracks" that were within the catastrophic crack propagation plane area.

Findings:
1. Sporadic crankshaft failures from the field contained hot cracks within the major failure plane on the thinnest counterweights.
2. All field failures to date involved the 4 thinnest counterweight sections.
3. Three of five failures were produced when 8 per hour of the crankshafts were dropped during handling in a two-day casting period (063 and 064).
4. Attempts to recreate hot cracks were not successful after the changes.

Conclusion:
1. Dropping hot crankshafts at the transfer point can create hot cracks.
2. Hot cracks sometimes occur on the thinner counterweights sections.
3. Hot cracks are contained within the cracking plane of failed cranks.
4. Catastrophic failures are related to hot cracks in the crankshafts.

Figure 6-3. Cracked Crankshaft Defect Location

There were only six failures, but each of them was on an inside counterweight that had the thinnest sections. We further identified each failure as possessing a hot crack at the fracture. After a thorough investigation, we determined that the fractures were the result of dropping one crankshaft upon another while they were being transferred while still red hot. The mishandling of the castings allowed some to drop more than the allowed distance from a shaker to a conveyor because there were missing system conveyor baskets. Some were cracked on the ends while others were cracked on the thin-sectioned counterweights, and there was damage only to some that were dropped.

Sometimes it's not possible to compare real-life good and flawed samples. In these cases, it is necessary to force a comparison sample. We took this approach when comparing leaking rear seals in engines that failed an air test. The relative frequency of a leak was 0.20% of the total production, which made it a difficult problem to solve. Unfortunately, all leaking seals were subject to mutilation when they were removed from the engine at the repair station. We removed samples for study. Since they were damaged upon removal, it was not possible to determine if they met the blueprint specifications after installation. They did meet the specifications before installation. However, when the faulty seals were replaced with new ones, the engines passed the air test.

As a result, we determined that there was a significant difference in the seals or in the installation method. To offset the impediments, we removed a few seals from engines that did not leak after the air test. Then we compared the good and bad seals, which generated two main clues that were instrumental in identifying the cause:

- Some seals had excess rubber from the assembly operation. This was considered to be foreign material (FM).

- Some seals appeared to lack grease at the seal lip surfaces.

There was clear separation of the visual data. The best seals (ones that didn't leak) had a better distribution of grease than did the worst seals. A defect diagram is often helpful in analyzing these kinds of problems.

The concept sheet shown in Figure 6-4 illustrates effects when components are removed from the established routing for part processing. In one case, a grinding operation created an oversized pin diameter that was unacceptable. The oversized parts were sequestered and reintroduced into the system at a rough grinding station rather than at the finished grinding station, as specified on the routing. The effect of the change might seem to be inconsequential but the resulting effects were disastrous. Because the grinding wheel moved and sensed the presence of the pin, it moved into the previously machined pins still in the grinding station and caused a taper on the adjacent pin end. Unfortunately, the pin was used in an automotive engine's piston assembly. Once it was assembled, the vibration during operation resulted in a

Chapter 6 | Generate Clues

catastrophic engine failure that stranded the motorists. Had the pin been placed into the finishing grinder, as specified in the routing, the taper would not have been created. Seemingly insignificant routing changes can therefore have a major impact and should not be overlooked.

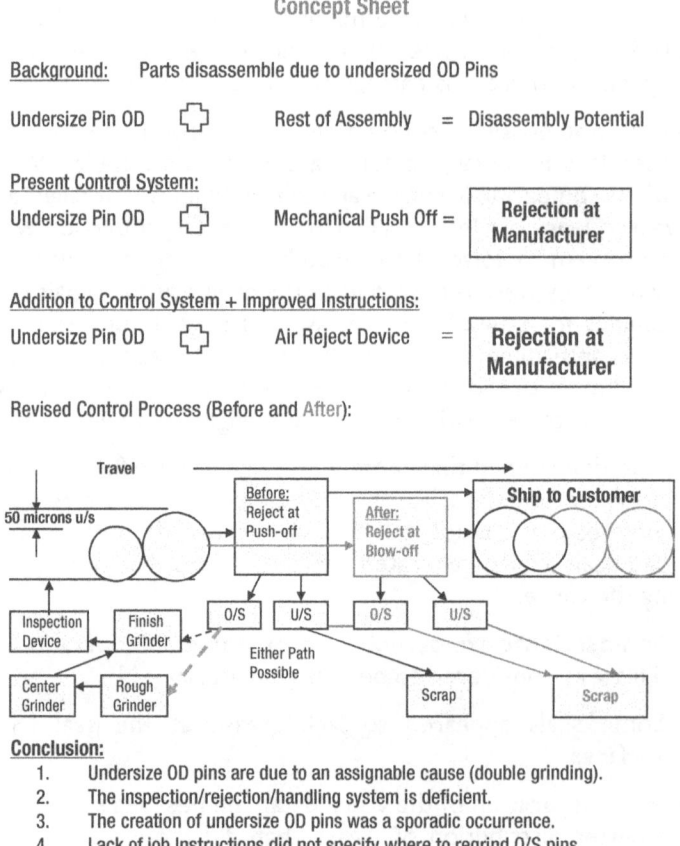

Figure 6-4. Example of Undersize Pin Concept Sheet

The sketch of the process in Figure 6-4 shows that there was an assignable cause to the engine failure problem. The "Before" path of a pin shown in black had undersize (U/S) or oversize (O/S) pins after a grinding operation, depicted on the left side of the sketch. The U/S pins were directed to the scrap bin, whereas the O/S pins were supposed to be placed into the finish grinder. However, the fault was that the oversize pins were placed back into the rough grinder rather than into the finish grinder. This caused the machine sensor to dive to a deeper level, which resulted in the tapering of the ground pin. The lack of posted consistent work instructions contributed to this problem.

The right side of the sketch shows the revised "After" path in bold; the procedures were changed so that all O/S and U/S pins are now directed to the scrap bin. A mechanical sorter was replaced with a more reliable air-actuated device to ensure nonconforming rejections.

Who would have thought that placing a pin in a wrong bin would eventually result in a catastrophic engine failure in a vehicle over 1,000 miles away? Who can estimate the amount of customer dissatisfaction or the lost dollars involved in trying to determine what caused the failure? More importantly, what can be done to prevent this event from happening again? Maintaining process controls should be a major focus, including using many level checks to ensure consistent work, and creating clear and concise job instructions.

After this problem was recognized and resolved, machine operators were trained as to the proper placement of oversize pins. One of the reasons that the problem occurred is because there were no instructions about what to do when machine operators found an oversize pin. We corrected the work instructions and control plan and retrained the operators. In addition, the quality personnel made adjustments in future DFMEA and PFMEA meetings and revised the control plan. Finally, the proper placement of undersize and oversize pins was added to the auditing check list to prevent recurrence.

Summary

This chapter provided a more micro view of the problem of generating clues. It took a deeper view, with examples of conditions that you may encounter. You can use job instructions, assessment plans, checks, audits, control plans, and other observed criteria to generate clues, as well as control an operation's output and quality. This chapter also delved deeper into the problem description and further differentiated it from the problem's cause.

The next chapter—Step 6—covers the use of innovative tools that you can use to solve problems.

CHAPTER

7

Choose and Use Analysis Tools

Step Six

This chapter is all about the analysis tools you can use to solve problems. Although there are many analysis tools available, each tool is fit for a particular purpose and provides clues in its own way. Some provide inductive information, for example, whereas others provide deductive information. Each provides a technique for identifying the differences between acceptable and unacceptable parts, processes, or actions, which can further reduce the time you need to solve problems.

Some of the most useful deductive tools have already been described. These tools require you to evaluate comparisons and generate ideas. The problem definition sheets discussed in Chapter 5, for example, show how visual examination allows you to make deductions about the flaw being observed.

This chapter examines innovative tools and methods that can generate a lot of information. Although they may require a more concentrated effort, they are still easily employed with a little practice. These tools are important because they allow you to move more directly into what may be causing the problem under study.

Six Prime Problem-Solving Tools

This section deals with the six most useful tools. I have provided examples of problems where these methods proved to be effective. These are the mini-power tools of problem solving:

- Basic analysis
- Trait peculiarities
- Sum of extremes
- Comparison of individuals, duos, and groups
- Fractional analysis (a method of using defect count data)
- Tests for clue generation or confirmation

Basic Analysis

As mentioned more than once, many problem solvers attempt to look at the defect and jump immediately to what their past experience indicates might be the cause. Unfortunately, this is not as productive as taking the time to perform a basic analysis correctly. Sometimes even the simplest problem can be troublesome. In most cases, the time required to resolve a problem is reduced when you take time to define it properly.

For example, issues with electrical components are some of the most difficult to evaluate, because they might not exhibit readily recognized characteristics. Pinched wires, bent connector pins, open circuits, shorts, and other assembly anomalies aren't readily identified because they are difficult to see. Here, as in most circumstances, basic analysis is vital to eliminating and preventing all problems, as well as to creating a body of corrective actions that can be used in the future.

Basic analysis can be used for problems well beyond the ones that occur in electrical assemblies. It can be applied to physical problems or conceptual problems even if the analyst has only limited knowledge of the system under review. Further, basic analysis focuses on the entire lifespan of the problem. To get to the root of the problem and ensure that it does not recur, you must ask of the problem condition:

- For what reason did the problem occur? (specific definition of the problem)
- For what reason wasn't the problem recognized and prevented?
- For what reason wasn't the problem identified when it occurred?

- For what reason wasn't the problem captured and contained?
- What corrective action can prevent the problem from recurring?

If the problem is defined correctly, each of the previous questions will provide significant actionable improvements for preventing the problem condition from happening again. Not only does the analysis aim to prevent recurrence, it initiates plans to recognize, identify, and contain similar problems with improvements to the quality process and efficiency. It is imperative to discover the reason for each of the questions before you can convincingly resolve a problem. A problem should be immediately discovered at its workstation and not pass to the next operation. It certainly should not be permitted to leave the operation or be delivered to the customer if it could be recognized at a prior operation. This is especially true when the customer has paid for the object.

Note There are many problem-solvers who attempt to ask "why" many times to pinpoint the problem cause. Unfortunately, the vast majority of the quality professionals who I have had experience with have not been properly trained to use that technique effectively. Therefore, without extensive training, this method appears to be inefficient in identifying the cause of the problem. Using one or more of the methods described in this chapter and the next will help you find the cause faster in most cases.

The purpose of a basic analysis is to observe faulty components as well as tooling, systems, and methods employed. The good news is that you have already learned the building blocks for a basic analysis.

First, you collect at least five samples of the units that are defective and five components that are satisfactory. Place them side by side and note any differences between the two distinct groups (bad versus good.) If you can't see any differences between any of the units in either group, inspect them with a magnifying glass or a microscope. If that does not yield results, use your measuring tools as applicable.

You can also check the components while recording the information on a problem definition sheet, as described in Chapter 5. Compare the fault present with the criteria on the defect scene characteristics table and the contrasts of two table (covered in Chapter 2). If applicable, construct a concept sheet that addresses the fault (see Chapter 3).

Finally, compare the fault to the criteria on the problem corrective action worksheet (see Chapter 1). Believe it or not, using these evaluative sheets will focus attention on the clues necessary to solve the problem in the most efficient manner. But these analyses focus attention only on the first step of the five presented at the beginning of the chapter—What is the problem and for what reason did it occur? This preliminary work will help you solve the problem.

The second question—For what reason wasn't the problem prevented?—may be discovered from the same corrective action worksheet. Generally, you'll find that training, routing, instruction, and procedural issues are not adequate or are not in compliance.

This question can be most effectively answered when you are defining the fault. For example, in the case of the cracked crankshafts discussed earlier, the hot crack was found to be due to an impact that resulted from parts being dropped from the handling system instead of being transferred to an overhead conveyor. There was also an unapproved change in the red-hot casting routing. The unapproved change was created by open spaces on the overhead carrier where a significant number of baskets had been removed for repair yet no replacements had been installed.

So the problem condition was caused by a broken conveyor stop mechanism and gaps in the handling system that allowed the castings to drop eight feet to a concrete floor. Since some castings fell upon others, the thinnest counterweights were damaged due to hot cracks in the formed castings.

To prevent the problem going forward, we had to establish an assessment to check the operation of the conveyor stop, ensure the presence of all baskets on the overhead conveyor, and ensure the lack of castings on the floor at the discharge end of the conveyor. Such checking operations must be performed at random times and independently by more than one person.

Why the problem wasn't identified, the third question, was due to a lack of awareness that handling abuse could cause damage. Assembly line personnel either didn't know or chose to ignore the operating instructions to stop the production conveyor and to notify the supervisor due to poor training:

1. Red-hot castings dropped on the floor could be damaged.
2. A conveyor stop was important to control the material flow.
3. Castings on the floor were not in the intended routing.
4. Missing baskets on the overhead conveyor would cause castings to fall.

Question four—For what reason weren't the damaged parts captured or contained?—was due again to a lack of awareness and training. A lack of supervision added to the problem.

Finally, the last question—What corrective action can prevent the problem from recurring?—should be fairly easy to answer once you've answered the first four. These were the actions we took to prevent it from happening again:

1. Immediately replaced damaged or missing baskets on conveyor.
2. Installed a gate to prevent parts from dropping to the floor.
3. Included basket and gate inspection for preventive maintenance.
4. Included basket and barrier operation in floor check listing criteria.
5. Revised the DFMEA, PFMEA, control plan, and work instructions.
6. Created a many-level assessment to check on these items.[1]
7. Conducted training of all the personnel involved in checking the operation.

As you can see in this example, there are many details to consider when evaluating a problem and taking ensuing actions. The basic analysis tool provided the insights required to solve the problem and prevent it from recurring.

The strongest clue was a hot crack that was identified in the crankshaft located adjacent to the crack initiation area. This, along with the casting date and the correlation with scrap records, resulted in the equipment malfunction findings. After defining the problem, identifying the cause is the most important step in the problem-solving process. The additional small clues of contrasts and individual identification differences increased the odds that we'd understand the cause. The use of the basic analysis tool has provided the insight to many past studies made in industry and service industries.

---

[1]The many-level checks are basically individual inspections, checks, or audits that are conducted by more than one person individually. The checks are comprised of individual items that may have been created over time that have been recognized to create problems with the product or process being reviewed. These checks can range from observations for the proper tools, methods, and materials being used in an assembly operation. They can also include inspection, safety, and other considerations that are not included here.

Trait Peculiarities

Many clues are available when you're conducting a study. The object of effective problem solving is to collect only useful clues. So, collecting the clues that cause the most differences is important in identifying the cause of the problem. This focus allows you to eliminate irrelevant variables. The relevant clues can be classified into different trait groupings. Further, these groupings can be separated into four major categories. These categories are:

- *Unusual differences*: When one rare trait appears or repeats in a pattern.
- *Piece trait differences*: Seemingly identical parts are found to be different.
- *Individual trait differences*: Differences found on a single piece.
- *Other traits*: Differences in material, machines, runs, workmanship, and so on.

The process of generating clues can be simplified if you can identify the variables that cause the largest differences. You can develop the means to choose the clues to be analyzed using a trait peculiarities drawing (see Figure 7-1). This drawing is a visual representation of the changes in the defect or condition that has been selected for comparison.

Industrial Problem Solving Simplified | 95

Trait Peculiarities Drawing

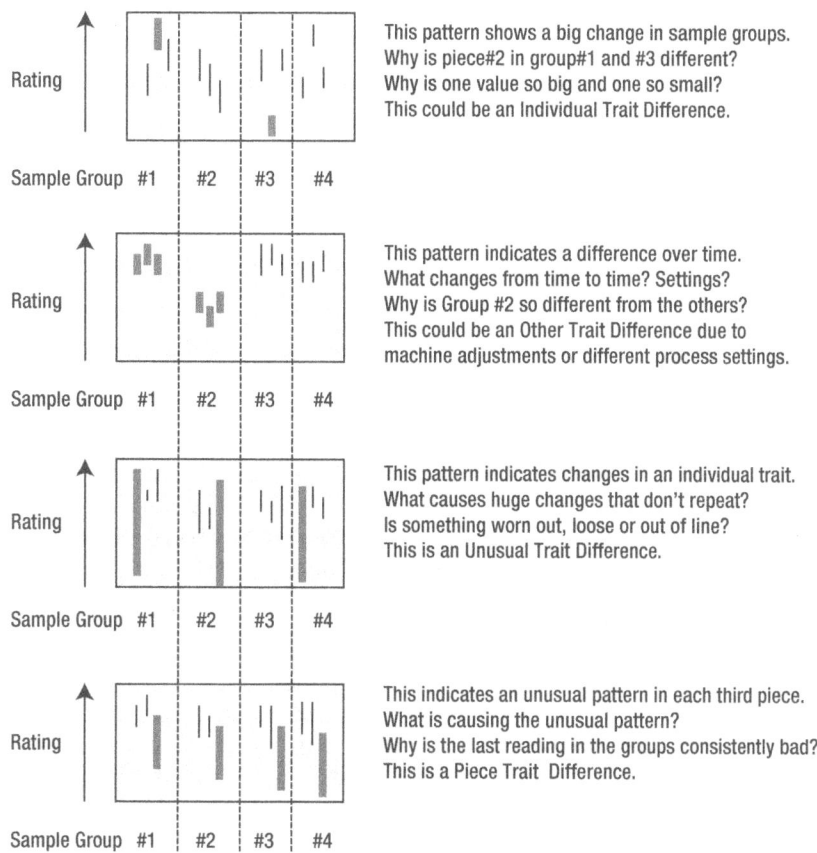

Figure 7-1. Generating Clues Based on Trait Peculiarities

■ **Tip** Almost everything you can see or test can be plotted and evaluated, including such things as hole size, true position of holes, alignment, pouring-basin weight, excess flash on components, the amount of shift, mismatch, roughness, strength of sewn bindings, pull strength, and more.

You can create trait peculiarity drawings by following these steps:

1. Collect three consecutive parts at least three times over three or more days.

2. Identify each part and measure the characteristic in question.

3. Record the individual identity and sequence of each part collected.
4. Attempt to capture the best and worst conditions being studied.
5. Rate the severity of the unfavorable condition for each part.
6. Plot the data on a drawing in the collected sequence.
7. Make comparisons and establish the major patterns.

The plots can be used to compare differences in characteristics or parts over different time periods.

Figure 7-1 illustrates this activity. For illustration purposes, only four groups of data are displayed for each condition described. The sheet displays these four different conditions as daily examples, with anomalies shown in thicker lines. The top display represents a plot for ground assembly pins. These plots were made by using the measurements obtained from ground assembly pins. The shaft diameters were measured twice. The measurements were made ninety degrees relative to each other and the data was plotted for each of the pins in each group for each of the four periods. This study led to a change in the process (discussed later).

In this case, the biggest difference in samples is shown in the first and third sections. The largest value in group 1 is shown much higher than the smallest value in group 3. This distance between the two samples was much greater than the length of each individual line, which represents the difference between the two measurements made for each diameter. So this is an individual trait difference, because one sample has a diameter much larger than a diameter measured in a following period.

In the two lower examples, the length of the individual lines could represent a change in diameters, lengths, or diameter tapers, depending upon the designation of the evaluator.

PLOT YOUR PROBLEM

The layouts in Figure 7-1 represent four individual daily readings. This illustrates one way that measurement data can be plotted and displayed.

Each sample group is designated for each of the four different days. The two uppermost comparisons show individuals or groups highlighted in thicker lines that pinpoint the two or more samples that showed the greatest variability, or difference, in the data.

Time differences are shown as the comparison to the readings on one day versus the other days. The vertical broken lines delineate the separation of one day from the next. Size differences are depicted as increasing when they are toward the top of the display and decreasing as they approach the bottom. This is the rating or measurement.

The distance of consideration is the greatest distance between highlighted samples that appear in the two uppermost example displays. The lower third and fourth groups show individual samples, which are not similar to other samples with the same sample batch of three. In this case, the difference is the magnitude between the longest and shortest sample measured on that day. However, because of their significance, these samples appear to be excessively bad and are highlighted because they are the most important consideration.

An individual trait difference is dominant on the top illustration because the largest variation is found to be between the biggest and the smallest piece. This type of data could be caused by parts machined on different machines or by different settings on the same machine because of operator tampering. Or it might indicate that a tool change was made.

The other trait difference, shown in the second display, could be caused by machine target adjustments or by different machines being used. This is why it is important to identify whether the process has more than one flow path. If it does, this pattern may reflect a difference in the machine used to form the shaft and could be a variation on machine tooling.

The next pattern shows unusual trait differences. This difference could be caused by a worn bearing, loose machining chuck, or another condition that prevents smooth machining along the entire length of the pin. If all of the samples are consistently deviant for within-piece variability, the machine may be out of alignment or affected by lack of lubrication, wear, or overheating.

The bottom pattern shows the piece trait differences pattern and it also requires investigation. The bottom pattern indicates an unusual display on each of the four days. This pattern does not necessarily indicate that the part is scrap, as it depends upon the designation when the comparison was created by the investigator.

In the next few pages, I'll describe two studies that had similar results. One case involved overheating batteries in a toy and the other involved variation in the size of bearing caps. A harmful temperature condition was created when the battery was installed backward in a toy. In the bearing cap case, presented later in the chapter, different dimensions were created in the castings that were produced using identical dies. This was because the sand was not compacted in the mold to the same density.

Toy Overheating Problem

A battery-operated toy designed to be used indoors and outdoors was overheating and spontaneously rupturing batteries in the drive mechanism. This sudden eruption caused severe damage to walls, rugs, draperies, and so on, due to uncontrolled carbon dispersion. Testing many units for temperature differences resulted in the pattern being exhibited when one of four batteries was inadvertently inserted backward into the battery compartment (see Figure 7-2). Once the problem was identified, we concluded that the other three batteries were operating in tandem to charge the fourth battery, which resulted in overheating. A redesign of the battery compartment to prevent batteries from being installed backward was a prime correction. In addition, the design (DFMEA) and process failure mode and effect analysis (PFMEA) were adjusted to prevent recurrence of this undesirable assembly operation.

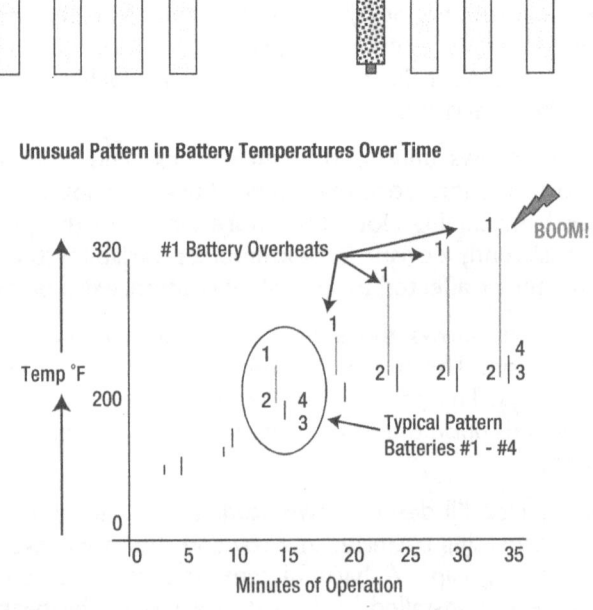

Figure 7-2. Battery Temperature Pattern

Figure 7-2 was generated in a quality control testing lab that conducted numerous experiments to determine whether the problem could be recreated. Luckily, one of the technicians installed a battery backward. If any of the four batteries was installed incorrectly, it was subject to an "interaction" due to the excess temperature and resultant failure. (Interactions are explained in Appendix B. Batteries these days can now only be installed correctly in most electrical devices to prevent these interactions.)

In summary, when an "other trait" pattern is present, it may indicate a shift or retargeting of the process, changed batches of materials, different environmental conditions, or different operator settings. Unusual trait difference changes may be due to worn equipment, bad bearings, lack of lubrication, machine adjustments, or outlier events. Some individual trait differences may be due to the use of two different machines making the same part or changing the machine target setting. These examples are not all-inclusive, as the process dictates the considerations that must be evaluated. The more the data can be digested, the easier it becomes to define the problem causing the defect. Therefore, each time you eliminate a variable, you're closer to identifying the problem. After you've identified the problem, you can pinpoint the cause.

Note One of the most important considerations in problem solving is to confirm that no unauthorized changes have been made to an approved process. These changes generally show up on a trait peculiarities drawing as another trait change. These changes can be attributable to changes in supplier, material, machining targets, manpower, methods, machines, environment, rework, or part routing. For example, an unauthorized reduction in lost foam pattern cure time caused a pinhole porosity flaw. The flaw was due to moisture on the pattern, which transformed into a gas defect when molten metal was introduced into a mold. There must be no unauthorized process changes made, no matter how slight. Your people must be trained to get approval for all internal and external changes in advance of making or approving them.

A constant in any manufacturing setting are finance managers or outside suppliers who propose material changes as a means to reduce costs. It is imperative that you be aware of any material or process changes to be made to any and all material that is provided to you. It is essential that your supplier and their suppliers be aware of this requirement. This means that your supplier and also their suppliers must not make any unauthorized change to any material or process that you have officially approved. Definitely be aware of any material or process changes or proposals that are contemplated. I cannot stress too much that these are the greatest flaw-creating situations that you will experience.

Note Unapproved changes to materials and processes could create more problems than just about any other cause—and their presence is difficult to discover.

Unfortunately, supplies from new suppliers can differ in chemistry, strength, machining, or quality capability, and you must fully test and approve new supplies before they're implemented.

For example, a new iron oxide supplier contributed to the creation of gas defects when it provided an iron oxide with a higher mix concentration of red iron oxide in place of the approved black iron oxide. Another failure occurred when a bearing pedestal formed of a substitute powdered metal failed under repeated loading after the supplier changed the material without authorization.

Then there are unauthorized procedural changes or poor training that must be discovered and prevented. An operator, for example, may intermittently apply too much pattern spray to a core box to lubricate a pattern. The excess spray will accumulate in the mold and cause gas porosity. Since the operator did not perform the soaking required during each machine cycle, the source of the problem was difficult to identify. This type of problem requires an enormous amount of investigation. We generated the clue using a trait peculiarities drawing. It plotted pinhole size and location in the casting versus differing amounts of pattern spray used in different mold sections, which made the issue obvious. Situations like this show that unclear job instructions and lack of training are frequent causes of quality problems.

Finally, suppliers do not always report the changes they make. When a problem is first being investigated, it is not unusual for suppliers to indicate that nothing has changed in the materials or processes at their location. Unfortunately, I have found that such assurances prove to be untrue about half the time, causing numerous problems. That's why the use of a many-leveled assessment to check for the use of consistent work is necessary. Remember that consistent work is the performance of a task in accordance with the specified requirements for using the materials, tools, and methods in compliance with the existing operating procedures and control plan.

Unfortunately, you must verify and re-verify compliance to the established requirements.

More Examples of Trait Differences Revealing Problems

A pin supplier finds that some of the coated pins that it supplies to another manufacturer have slight dents and dings that make them unacceptable. In an attempt to salvage the parts and to keep supplies flowing, the supplier decides to polish or buff the pins before shipment to the customer. The supplier finds that a hand grinder that contained an aluminum-based polishing wheel does a good job of buffing the pins to make them look acceptable. Unfortunately, the pins were then contaminated with aluminum oxide particulate from the grinding wheel, which can destroy an engine due to tight tolerances. This condition may display as another trait difference on a plot. These differences can show up due to material, machines, runs, workmanship, and so on. Hand buffing these pins would have shown up as a finish difference between the sample parts.

Aluminum oxide polishing cloths and similar type products have been widely used acceptably in many service and manufacturing applications. Unfortunately, they can also be misused and cause extreme damage when applied to engine components that will be assembled. These same aluminum oxide or carbide particles can enter a close tolerance system and cause interference and distress to the mating components that are subject to oscillation and friction. Contaminants and foreign materials (FM) can cause imperfections in the process that will result in failure. Maintenance personnel may have these products in their toolboxes even though the company has banned them. When detrimental products are banned, it is necessary to thoroughly notify and instruct all personnel of the importance and purpose of such bans.

As shown on the concept sheet in Chapter 6 (see Figure 6-4), pins that have an oversized diameter after processing and are placed back into the rough grinder instead of into the finish grinder as stated in the approved routing can cause costly failures. Because they are smaller than the other rough stock components being supplied to the rough grinder, the grinding wheel can cause a taper to be introduced into the pin, which can create a piston disassembly problem in the field. A trait peculiarity drawing defined the differences in the diameters on either end of the pin as the suspect condition. This was a "piece trait" difference, as shown on the bottom drawing in Figure 7-1.

The text in black on Figure 6-4 shows the path of the ground parts before the problem was encountered. The lighter lines and text indicates the flow path after the changes were made to correct the process. The problem was created in the before flow path section, as indicated by the dashed line, when the oversized parts were placed into the rough grinder rather than the finish grinder. This caused a regrinding taper that was unacceptable. The after flow path shows that the oversized and undersize pins are to be placed into the scrap bin. Note that there were no undersized or tapered pins getting sent to the customer after the change had been made. So even a check of the final product can be evaluated using the part trait method.

Look at the Forces Involved

Analyzing how energy is applied can play an important role in solving some problems. Questions to be answered will point you toward the correct solution path. For example, did only one of many parts break? Where did the break start? Is the break in the same location and pattern on all the broken parts? Is there any kinetic or potential energy that could have impacted the part? Did the part fail because of impact? Did the part fail because it was too weak or because the force applied to it was excessive?

This could be shown as an unusual trait difference on a plot if broken parts did not all come from the same pattern serial. It could also appear as a piece trait difference if a broken part was always produced by the same pattern serial as the other broken parts.

Here's an example. A toy plastic horseshoe cracked after it was tossed because it contained voids. The voids caused a weakness that was exposed by the force of impact (see Figure 7-3). Eventually, we discovered that the operator had changed the machine cycle of the plastic bath, which prevented full mold filling, but let's back up and see how we got there.

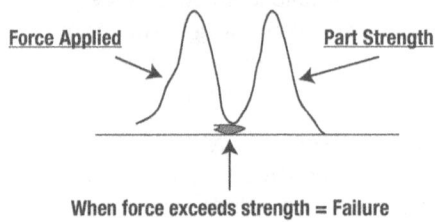

Figure 7-3. Force Applied Versus Part Strength Diagram

First, we had to determine whether excessive force was applied by someone throwing the horseshoe, or whether the part lacked adequate strength to fulfill its form and function. That's one of the first questions you should ask when energy (damage) is suspect. Since the horseshoe was designed to be tossed in the game, the strength of the toy was more suspect than the impact force.

We decided to use the pull strength as an indicator that could be used in a trait peculiarity chart plot for evaluation. Plotting the pull strength of different horseshoes from within the mold containing four inserts provided clues. The shoes from mold 3 were always weaker than the other molds. This was a shoe-to-shoe difference. If each shoe showed the same relative strength, but the average strength differed over time, it could indicate a material piece trait difference. If samples from two different machines showed different average strengths, it could be an individual trait difference based on production machines.

The force to break a horseshoe might differ from shoe to shoe as a result of different flashing removal operations. They could differ from one time to another due to different batch amounts of regrind material use. Or if one area cracked differently from another, the trimming process to remove the gates and runner might be the culprit. From the data obtained from the pull strength tests, we determined that the weakest horseshoe was always the one that was made from pattern serial 3. The plot revealed differences between this part and the other three that were made in the same mold. Because these were plastic parts, they were sectioned, which revealed that number 3 parts had excess voids that affected their strength. We changed the plastic flow pattern to correct the condition.

The flexibility of using trait peculiarity studies to collect relevant clues is unlimited. Once you plot the most compelling data, you analyze the comparison patterns to find the clue.

Bearing Cap Case: Part Attribute Analysis

Here's a similar problem to the horseshoe flaw. A machining line was experiencing downtime because some bearing caps from the supplier would jam in the grinder-processing equipment. This was a chronic problem that was worse on some days than others. We checked the machining equipment and found it to be acceptable. By measuring, we found that the problem always seemed to be with the iron casting 1, 2, 5, and 6. Measuring the individual caps showed that castings 3 and 4 were thinner than the others. This measurement was the comparison of the widths of the bearing caps that were made for each bearing cap in each position in the mold. Since there were six bearing cap pattern serials in each mold, each casting was represented by a number, 1 thru 6.

There were six bearing caps made within the same mold (see Figure 7-4). Each of the pattern serials conformed to measurement specifications and were identified as 1 through 6, as shown in Figure 7-4. We measured six caps in each of three molds at three different times. Each of the groups contained the range of bearing cap widths measured in each of the three molds.

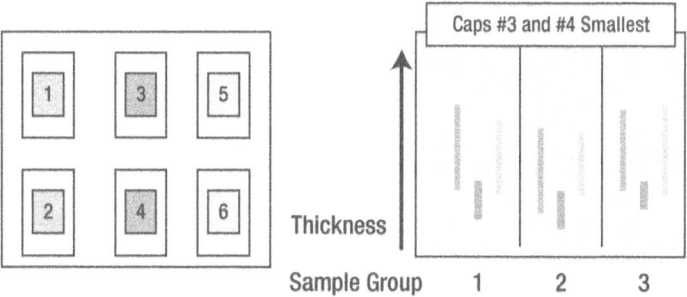

Figure 7-4. Basic Analysis of Bearing Cap Variables

We found that castings 3 and 4 (the lowest bar in each group) always had the smallest width. Since the castings were made from serials that were virtually identical, the finding was troublesome. However, the size differences were not consistent casting to casting. There appeared to be swelling at the top of the 5 bearing cap, whereas there appeared to be more prevalent swelling at the bottom of 2 bearing cap. The picture was still clouded but valuable clues had been obtained.

We decided to gather data from the mold because a concept diagram indicated that the cause of the binding of the production machine was either due to excess flash on the sides of the castings or a condition of mold wall movement. (*Mold wall movement* is a condition that results when molten iron is poured into green sand molds. The heat of the iron drives the moisture in the sand away from the interface and in effect allows the mold wall to move outward. This can allow the casting to become larger than intended.)

We selected molding properties as the suspect condition because the bearing caps were not considered smart enough to know where to swell unless acted upon by an assignable cause. We took mold-hardness readings and plotted the data, as shown in Figure 7-5. The readings were taken three times at each side and three times closer to the center.

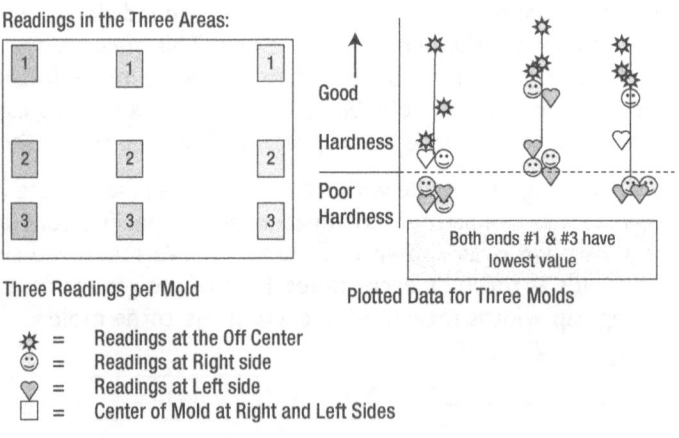

Figure 7-5. Mold Hardness Data Plotted for Three Molds

The first group in the image on the right shows the results from the first group of three molds. The second and third groups further to the right showed almost identical patterns. The findings from the part attribute drawing indicate the following:

1. The center of the mold always had the highest mold hardness.
2. The two central readings on either end were harder than the corners.
3. The mold hardness in the corners was low and not to requirements.

The cause of the insufficient mold hardness was found to be insufficient sand drops from the sand hopper and insufficient squeeze of the sand on the mold pattern to create the mold cavity. The solution was rather easy; we changed the sand gate height and added tucking pads to the machine to help squeeze the sand more firmly in the corners. This helped prevent mold wall movement and the resulting swelling distortion.

A trial run after the corrections were made showed all hardness readings meeting requirements. The new samples passed through the processing machining without causing a jam-up.

This didn't end the project. Checking the mold hardness and recording the processing data became part of the work instructions. This type of activity helps to prevent the similar problem from recurring.

In addition, we added the assessment of mold hardness to a many-level check sheet to help preserve the gains that had been made. These two items of consistent work and assessments are explained in later sections.

SOMETIMES IT'S THE CUSTOMER

Sometimes the customer or user causes the problem. A toy manufacturer of electrically driven vehicles received reports of home fires caused by one of their vehicles, when they were being recharged. Fires were reported to have started on porches, in garages, and alongside the houses.

Because there was no significant information or testing to indicate this type of problem, we had to conduct an autopsy on the failed vehicles. Upon closer investigation of the melted plastic masses, it was discovered that the customers had inserted foreign material into the fuse slot on the battery on each of the vehicles.

Users were attempting to bypass the electrical safety fuse system. Pieces of washers, thick wire, paper clips, and other unidentifiable metal material was installed to replace blown safety fuses. If you can't prevent a product failure, you must provide clear instructions to the customer.

The next section explains tools that display and document collected data in pictorial form. The tools also provide a project summary so that you can easily identify actual current status of the visual clues.

Sum of Extremes Test

The sum of extremes test is based on the fact that there are only 20 ways that two groups of three items can be arranged numerically or combined without duplication or ties. For example, consider three good and three bad items. Arrange them by groups, and you'll get 20 different combinations.

Chapter 7 | Choose and Use Analysis Tools

If you have doubts about this, you can challenge yourself by attempting to prove it incorrect. However, you may want to look in Appendix E for an example that verifies this factual statement. (Or, you can wait until later in this chapter, where the 20 different combinations are displayed.)

Similarly, there are only 70 ways that two groups of four can be arranged without duplication or ties, and 252 ways that two groups of five can be arranged without duplication or ties. Experience has shown that it is creditable and convenient to use five good and five bad samples for comparison. Generally, the differences can be observed if the objects are compared or studied individually. As mentioned, the sample of five good and five bad also leads to other comparisons, verifications, and tests that have proven to be effective for problem resolution. Comparing five of both conditions gives good results.

How effective are they? In taking a sample of five good and five bad, there are only eight distinct combinations of the 252 combinations where the extremities are ranked, counted, and not mixed, and they result in tally of extremes of seven or more. As an example, let's use the combination of three good and three bad for simplification. This same method is applicable for all other combinations, including four good and four bad and/or five good and five bad. Assume that the good values have a lower leak rate and are measured to have a low number. The bad samples have a higher leak rate and a higher number. The first comparison that shows the separation of the good from the bad might be as follows:

	Better	Better	Better	**Inferior**	**Inferior**	**Inferior**
Test value	1	2	4	7	9	10

There is clear separation of the good and the bad based on the test score shown. This row of data comparisons would have an extremes count of three good and three bad, which is a value of 6, because the good and bad values are clearly separated. When data overlaps or is tied it cannot be counted into the extremes count value. This is shown with different data as a tie in values:

	Better	Better	**Inferior**	Better	**Inferior**	**Inferior**
Test value	1	2	4	4	7	10

The extremes test result in two good and two bad is a value of 4. Because the values are tied or overlapping, they cannot be counted in the extremes test value. This will be more apparent in the examples given later in this chapter. The basis for the combinations and examples of the values that do not overlap will be fully described and explained next.

Industrial Problem Solving Simplified

When each of the arrangements in the preferred method is compared and the data is arranged by sequential value, the addition of the acceptable (good) on one end and the unacceptable (bad) on the other end results in a value of 7 or more.

Before you get into specifics of how the comparison of five good and five bad works, it is necessary to understand how the comparison is to be conducted. Done right, you can achieve greater than 95% confidence in results.

First, the two data conditions cannot overlap to be used in the sum of extremes test. You must discount any tie value between the good and bad and arrange the most disadvantageous position in the ranking. (This may appear confusing at first, but examples will provide clarity.) G stands for good values and R stands for proposed replacement values in the drawings. If the values tie, they are not shown in bold and they are not counted for the required extremes total of 7.

Using five samples from each group reduces the costs of the study and still delivers at least 95% confidence in the results if sum of extremes is 7 or more.[2]

Caution Because of chance, you cannot be 100% certain of your result. However, being 95% confident is much better than being 50% certain. Therefore, after each study, always attempt to verify the results with trial runs or statistical tests. You can do this by running five samples of the original material and five samples of the replacement material and comparing the results. If all five duo comparisons fall in the same classification, you can be more than 95% confident that you can make the right decision with the proposed material change. This is based on the assumption that each evaluation gives a 50% chance for success. If all five duo comparisons indicate uniformity and acceptability, you can be 95% confident that the change can be made or is valid:

(0.5 x 0.5 x 0.5 x 0.5 x 0.5) x 100 = 3.125. Then, 100% − 3.125% = > 95 %

---

[2]This is verified by the calculation of 8 divided by 252, which equals 0.031746. The 8 is from the total acceptable extremes combination of five good and five bad. The 252 is the total number of ways that the two samples can be listed without duplication or overlap. Therefore, the total of (1.0 − 0.031746) x 100 equals 96.8254%. That shows 95%+ confidence and that there is a significant difference between groups. There are only eight ways to get a total of seven or more extremes with a comparison of two groups of five.

Sum of Extremes Analysis

Figure 7-6 shows examples of analysis results. They involve the comparison of samples from both a current (G) and a proposed replacement (R) sample population. The sketches show any overlap in the population characteristic being measured. If there is complete separation of the data groups, they are likely from different populations. The separation of the recorded values from the R (Replacement) or the G (Current) populations often indicates that there may be a significant statistical difference even when some of the data overlaps.

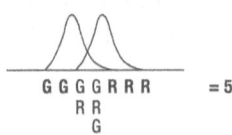

Acceptable Extremes:
With 4 Gs lower and 3 Rs higher
Cumulative extremes = 7
R may be better than G.

Sample R has higher values

G G G G R R R R   = 7
        G
        R

Unacceptable Extremes:
With 2 Gs lower an 3 Rs higher
Cumulative extremes = 5
R is not better than G.

There are too many overlapping points

G G G G R R R   = 5
      R R
       G

Figure 7-6. Sum of Extremes Explanation and Examples

For example, the upper left sketch indicates that with a sample of five G and five R measurements there is over 95% confidence that there is an acceptable difference between the two groups. (Confidence is a measure of sureness.) The total sum of extremes meets requirements of 7 or more with no overlapping data values. From the diagram, you can see 4 Gs + 3 Rs = 7 cumulative extremes. If the measurement values overlap, as shown in the right-side sketch, where only three Rs are larger than two Gs, there is no confidence of any statistical difference between them. The sum of extremes is 3 Rs + 2 Gs = 5, which is not acceptable. There must be 7 or more in the two groups of five.

A tally of 6 or less indicates that the measurement may be from similar or different populations or may have happened by chance alone. This chance occurrence is minimized by using five of each sample. Depending on the setup for the test, the data could indicate that there is a lower rate of defectives with the R condition or it could mean that the current product (G) is better than the proposed replacement part (R). (It depends on the original definition used.)

A sample follows in which five current and five proposed pipes were tested to verify the quality of a proposed supplier's product. Current pipes were tested as were five proposed pipes. The results are shown in Figure 7-7.

Sum of Extremes for Comparison of Two Samples

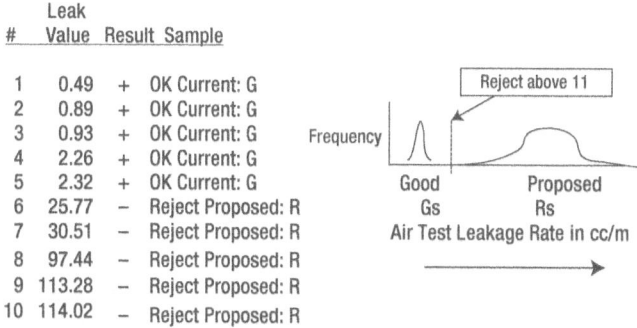

```
     Leak
  #  Value  Result  Sample
  1  0.49   +       OK Current: G
  2  0.89   +       OK Current: G
  3  0.93   +       OK Current: G
  4  2.26   +       OK Current: G
  5  2.32   +       OK Current: G
  6  25.77  -       Reject Proposed: R
  7  30.51  -       Reject Proposed: R
  8  97.44  -       Reject Proposed: R
  9  113.28 -       Reject Proposed: R
 10  114.02 -       Reject Proposed: R
```

Figure 7-7. Sum of Extremes for Comparison of Two Samples

We applied a sum of extremes test to verify separation of the five current and five proposed pipe assemblies with an air test. Although tested at random, the results of leaking in cubic centimeters per minute (cc/m) are displayed in rank form. Each of the good pipes tested much lower than the proposed pipes on an air test and there was clear separation of the data. All of the current G samples passed the test with lower values than the proposed R samples. Therefore, this test had an acceptable sum of extremes value of 10, which exceeded 7.s

Sum of extremes = 10 (5 G + 5 R). Groups were separated with no data overlap. This meant that the proposed pipes had greater leakage and should not be accepted or used as replacements for the current pipes.

The results of the air test showed that each of the proposed pipe assemblies leaked more than the allowable specification and more than the current pipe assemblies. The comparison indicated that the minimum count of 7 was achieved for statistical significance when the test consisted of five good and five proposed samples. Since the cumulative count was 10 (and therefore greater than 7), there is more than 95% confidence that the two groups are statistically different. After measurement, the critical difference was found to be in the individual fuel pipe diameter dimensions.

For those inexperienced with sum of extremes testing, it is advisable to test a minimum of five samples of each condition. If five random units of each condition are used and there is not more than three data points equal and/or overlapped, the evaluator can be certain that there is at least 95% confidence that the two samples groups are indeed different. This amount of samples allows you to possess the data for all of the tests and evaluations used in the analysis. I recommend that you save the acceptable samples from the study to provide background information for sample reviews if required. (Remember that confidence does not equal certainty.)

Comparison of Individuals in Duos (Sets of Two Units)

There is another valuable tool that you can use with most manufacturing problems, called the comparison of duos. A duo is a matched pair of individual samples that are scrutinized for differences. This tool can be used to compare a leaking fuel pipe to one that doesn't leak. It might be used to compare a present part to a replacement part to determine which is better. Consider the following example, which can make the concept clearer.

A leaking fuel rail (pipe) problem was discovered on an engine assembly line. Production stopped and an immediate containment was issued to capture the suspect engines. We started an evaluation by constructing a part attribute analysis that measured the circularity (roundness) of the fuel pipes. Measurements were made at 90-degree increments when it was suspected that some appeared to be out of round. The leaking fuel tubes did not pass the air test and appeared to be from the same supplier but from a different shipment. These fuel lines (identified as B) were more oval and not as round as the current acceptable fuel pipes (identified as A). The biggest difference that appeared on a drawing analysis was the difference in circularity of the parts, which approached 0.002 of an inch. Also, the presence of a groove in the weld bead area reduced the effective minimum diameter by the depth of the groove on the leaking fuel lines.

To verify the finding, we inserted a mixed series of identified fuel rails randomly into the production line. Each time one of the units arrived at the test station, it was marked **O** if it was rejected by the air test and by the starburst if it was acceptable. Figure 7-8 shows the results.

Comparison of Individuals in Duos (Sets of Two)

Trait vs. Duo Sample	Duo #1 Good A	Duo #1 Bad B	Duo #2 Good A	Duo #2 Bad B	Duo #3 Good A	Duo #3 Bad B	Duo #4 Good A	Duo #4 Bad B	Duo #5 Good A	Duo #5 Bad B
Minimum Diameter	☼	0	☼	0	☼	0	☼	0	☼	0
Diameter Difference	☼	0	☼	0	☼	0	☼	0	☼	0

Figure 7-8. Comparison of Individuals in Duos (Diameters)

■ **Tip** You can use any symbols to differentiate between good and bad results—plus or minus signs, directional arrows, Xs and Os, or what have you. Use whatever is most meaningful to you and that others can interpret and recognize quickly and accurately.

Visually comparing the duos will show separation of the observable data as provided by the ranking indication. This allowed us to evaluate the current and proposed components as a pair. The minimum diameter of each bad part was less than the minimum diameter of a matching good part, as shown in Figure 7-8. (Arrows can be used to indicate greater or lesser trait values.) In addition, the difference of the diameters measured at 90 degrees from each other ("diameter difference" in the figure) showed that the acceptable parts had less out-of-round conditions than did the unacceptable parts. This data showed that there was clear separation of the test data that gave a degree of confidence that the new shipment characteristics were different and not an equal to the samples in the existing group.

Figure 7-8 shows how five good and five bad fuel pipes compare. They were compared on a good to bad basis five times with one good sample and one bad sample in each duo comparison. Even though this comparison is not as thorough as the sum of extremes test, it does allow you to make preliminary visual findings. In this case, it showed that there was a significant difference between the groups under study. Attribute comparisons are not as thorough or as specific as numerical contrasts, but they are still useful.

If there are clear differences between the samples in each duo, then there will be a pattern of the comparison, where all of the good or all of the bad samples contain a measurable characteristic.

In Figure 7-8, the minimum diameter and diameter difference is different in the good (A) and the bad (B) fuel line samples. This means that they appear to be significantly different. However, Figure 7-9 shows that there is no difference in the lengths of the fuel pipes, as all the display marks show equal or nondiscernable values. This means that there is no reason to suspect that they are significantly different. The surface finish data, on the other hand, suggests that there is dissimilarity between the good and bad parts in some duo comparisons. Results are mixed.

	Duo #1		Duo #2		Duo #3		Duo #4		Duo #5	
Trait vs. Duo Sample	A	B	A	B	A	B	A	B	A	B
Fuel Pipe Length	*	*	*	*	*	*	*	*	*	*
Surface Finish	*	0↑	*	*	0↑	0↑	*	0↑	*	*

Figure 7-9. Comparison of Duos (Length and Finish)

If measurement values of good and supposedly bad samples are equal, as shown in the fuel pump length example, this means the parts are similar and no comparison of this trait can be made. Surface finish results show that the two samples overlap and that they may be significantly different when compared. Or at least the populations overlap.

Comparing the two units in each duo shows that minimum diameter and roundness of tubing are suspect in Figure 7-8. This can be noted by the visual signal that all the indicators in the row are consistent for each set of duos but not for each individual in the duo set. The proposed (bad) unit always had a smaller minimum diameter than the current (good) unit. Also, the diameter difference of the proposed (bad) unit was always more out-of-round than its duo matching current (good) unit. The consistency of the indicator in Figure 7-9 for fuel pipe length indicates that there is no significant difference between the two groups.

Earlier in this section, you learned that using five of each sample can provide more than 95% confidence in the results. However, there is a caveat to that assumption. When comparing duos, it's possible for more than three values of the proposed replacement material or process to be more acceptable than the materials or process to which it's being compared. In this event, there is no confidence, because the necessary seven minimum cumulative extremes are not available. Figure 7-10 shows an example of when a maximum pull strength in pounds is required and duos are being compared.

Material Being Used:		Material Proposed:		Extremes Not Present
G	44	R	34	Cumulative extremes = 2
G	32	R	30	
G	31	R	28	(One G + One R)
G	26	R	25	
G	24	R	23	GRGRGRGRGR

Figure 7-10. Sum of Extremes Example for Materials

Each of the G values is stronger than the matching R values in each duo. However, value R = 34 is greater than G = 32 and G = 31. R = 30 is greater than G = 26 and G = 24. R = 28 is greater than G = 26. Also, R = 25 is greater than G = 24. So the cumulative extremes is calculated by the left G and the most right extreme R. All the rest of the G and R values overlap. This gives a sum of extremes value of 1 + 1 = 2. So be aware that even two samples of five each can have similar overlap and therefore they will not meet the sum of extremes test value of 7 minimum extremes. (Tests of extremes give confidence in the result, not in the certainty!)

In most circumstances, the comparison of five duos will not result in this anomaly and will provide a good clue for further evaluation. The presence of the individual data for comparison will point to a potential cause. In most cases, the patterns exhibited by the data will be random and will not show a pattern, as it should not happen by chance alone. (However, you now know to look for this anomaly.)

There is a way to generate a significant difference between an acceptable sample of three and a test sample of three, as shown in Figure 7-11. Consider a sample currently being used as "better" and a proposed sample being classified as "inferior." Since we know the acceptance of the current sample, we should always test the proposed replacement sample as being suspect because it has not yet been certified or approved for use. Even if it is designated as inferior for the trial, it may be a better material if it surpasses the current material. We want to know which is the optimal material.

1.	Better	Better	Better	Inferior	Inferior	Inferior
2.	Better	Better	Inferior	Better	Inferior	Inferior
3.	Better	Better	Inferior	Inferior	Better	Inferior
4.	Better	Better	Inferior	Inferior	Inferior	Better
5.	Better	Inferior	Better	Better	Inferior	Inferior
6.	Better	Inferior	Better	Inferior	Better	Inferior
7.	Better	Inferior	Better	Inferior	Inferior	Better
8.	Better	Inferior	Inferior	Better	Better	Inferior
9.	Better	Inferior	Inferior	Better	Inferior	Better
10.	Better	Inferior	Inferior	Inferior	Better	Better
11.	Inferior	Better	Better	Better	Inferior	Inferior
12.	Inferior	Better	Better	Inferior	Inferior	Better
13.	Inferior	Better	Better	Inferior	Better	Inferior
14.	Inferior	Better	Inferior	Better	Better	Inferior
15.	Inferior	Better	Inferior	Better	Inferior	Better
16.	Inferior	Better	Inferior	Inferior	Better	Better
17.	Inferior	Inferior	Better	Better	Better	Inferior
18.	Inferior	Inferior	Better	Better	Inferior	Better
19.	Inferior	Inferior	Better	Inferior	Better	Better
20.	Inferior	Inferior	Inferior	Better	Better	Better

Figure 7-11. Sample of 20 Matrix (Better Versus Inferior)

Suppose for example that we designate "better" for the material currently being used and "inferior" for the proposed material. We can then make our test and determine the sequence of outcomes, as shown in Figure 7-11. Only in condition 1 of 20 shown below do the three "better" results fall to the left of the three "inferior" results. There is clear separation between the two groups that show better to the left of inferior. Depending on the test requirement, case 1 could be the better condition if lower test values are required. Or, case 20 could indicates that a new method or material, which was called "inferior" for example, actually has a lower test output than the existing "better" method. This would indicate that the replacement material should be considered for approval and use if it meets the desired test criteria. The following example provides an explanation.

Suppose you want to show the number of ways that two groups of three items (six items total) can be combined when taken three at a time in each group. Recall that there can be 20 different ways in which the results of the test can be ranked or described. These conditions depending on test definitions are shown in Figure 7-11. (I provided this sample in place of a sample of five from each of two groups, because it would have necessitated the listing of over 252 different arrangement combinations.)

Condition 1 shows that all the better parts are ranked lower. In this case, each of the inferior parts had a higher (or more unacceptable) test score than each of the better parts. Since they ranked completely different, there was no data overlap and there was clear separation of the two group's measurements. So if only one of 20 combinations can produce the results in condition 1, there is only a 5% chance that it can occur by chance alone: $1/20 = 0.05$, or 5%.

Condition 20, on the other hand, indicates that the inferior samples ranked lower than the better samples as a result of an imaginary test score. So again, if only one of 20 combinations can produce these results in Condition 20, there is only a 5% chance that it can occur by chance alone. That is $1/20 = 0.05$ or 5%.

You may be inclined to conclude that there is a 95% chance that your comparison shows that a significant difference between them is present: 100% - 5% = 95%. The reason that you cannot be certain about this is that there is a 5% chance that the result you obtained that indicated 95% confidence will be incorrect 5% of the time. Now I know that that is confusing, and it's why I've attempted to write these chapters without requiring the understanding of statistical theory or statistical calculations.

Let's look at another example, which ranks, separates, and compares samples to determine which product should be used. Two different conditions of fuel pipes are compared while investigating a leak problem. We compared a round diameter pipe and what appeared to be an oval diameter pipe. These were designated C and L, respectively. We made measurements and each of the C pipes had a lower or nonexistent leak rate than did the L pipes, with the oval diameter. (Be advised that the test criteria was excessive in order to induce leaks for this test. Leaking fuel pipes would never be allowed to be used.)

These values are shown as pass and fail for each of the six pipes. You can see the separation in the distributions, as there is no overlap in the values between the C and L pipes. If the data is ranked by leak value and then arranged, it results in a "C C C L L L" pattern.

From the results, we can be 95% confident that the round fuel pipes are significantly different and better for this application. They are much less likely to leak.

Industrial Problem Solving Simplified 115

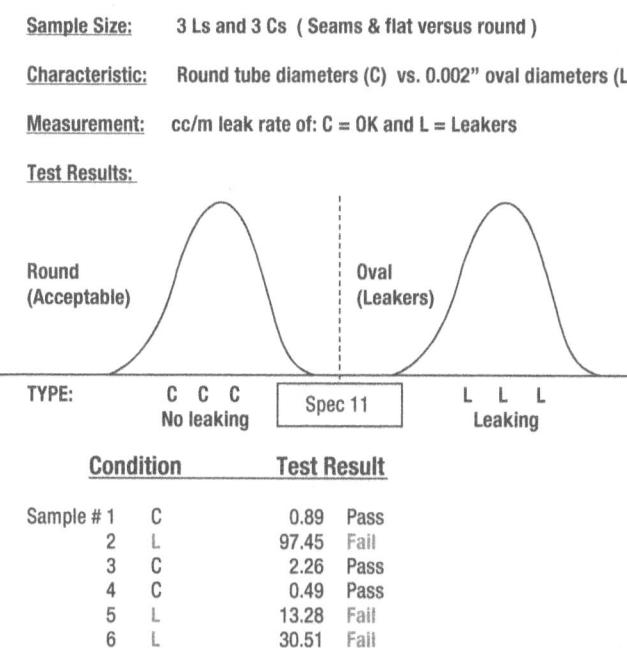

Figure 7-12. Better Versus Inferior Fuel Pipe Leak Test

Comparison of Individuals in Groups

If you're unable to separate your samples into good and bad groups with certainty, you can use a process that compares data by groups as the evaluative tool. The test actually employs a sum of extremes technique. It compares the rank of data generated by the individuals in each of the two groups as they compare to the individual, ranked values of both groups combined.

Say for example that the data in Figure 7-13 has been collected for a single trait.

Chapter 7 | Choose and Use Analysis Tools

Good Unit Values: 2, 1, 2, 1, 2.
Bad Unit Values: 3, 3, 4, 3, 3

1) Assemble the units in table form.   2) Place in rank order.

Data		Value	Good or Bad Unit
2		1	Good
1		1	Good
2		2	Good
1		2	Good = 5
2		2	Good
3		3	Bad
3		3	Bad
4		3	Bad = 5
3		3	Bad
3		4	Bad

Figure 7-13. Comparison of Individuals in Groups

Since no bad value is less than or equal to a good value, the extremes sum is 10 additive extremes. Remember, counts of difference of 7 or more are required to give 95% confidence. This shows clear separation and provides 95% confidence that there is a significant difference. The count is obtained by adding the number of extremes from either side that are not equal or that overlap. Since all five good samples have lower values than the five bad samples, the sum of acceptable extremes is equal to 10 (5 + 5 = 10). (Which means that there is a 95% chance that there is a significant difference between the two groups.)

The next example illustrates the method of determining the sum of extremes when there is data overlap. (See Figure 7-14.) This additional example is provided for clarity to show you how to handle tied scores.

Good Unit Values: 2, 1, 2, 1, 3.
Bad Unit Values: 3, 3, 4, 3, 3

1) Assemble the units in table form.   2) Place in rank order.

Value Unit			Value	Good or Bad
2			1	Good
1			1	Good
2			2	Good = 4
1			2	Good
3	The Good and		3	Bad
3	Bad Values	These	3	Bad
3	Are Equal.	Values	3	Bad
3	Don't Count Ties.	Overlap	3	Bad
3			3	Good
4			4	Bad = 1

Figure 7-14. Comparison of Individuals in Groups

Industrial Problem Solving Simplified

Remember, a sum of extremes of 7 or more must be present for 95% confidence. The tally here is 5 because ties cancel each other out.

Four values at the top and one at the bottom do not equal the required extremes value, so we can't conclude that there is a real difference.

The reason that the count at the bottom is one is because the good value of three is equal to the bad value of three, which was measured for one of four individual bad samples. This prevents it from taking a position in the tied ranking. It could be in the fifth or the ninth position Therefore, it should be considered to be in the most detrimental position, which is the ninth position from the top. Since there was one value of 4, it is the only unit measured at the bottom for an extreme. Both of the extremes of 4 and 1 mean a cumulative sum of five extremes.

Note that the confidence level is below 95% if more than one trait from the sample is being compared, due to the fact that at least 5% error uncertainty comes into play. (If you simultaneously check 100 traits for 95% confidence, you may find five of the 100 comparison results could be incorrect due to the 5% uncertainty in confidence. Inaccurate results could be higher or lower, so be aware of this weakness when designing tests).

Group Data Applications

Sometimes when you're studying a complex process, it is not readily possible to recognize the significant information. Consequently, it may be necessary to create a data sheet and collect all the peripheral information.

Figure 7-15 shows three tables that contain the peripheral data from an aluminum void study. Because the flaw under study was the porosity and the resulting rating, we used the quality of the final product as the measuring indicator. Because we were not certain what caused the porosity issue, we gathered data from six different variables. We collected the data and placed it in a table, as shown in the top-left side of Figure 7-15. We developed a rating system and classified the porosity in the casting for evaluation. This was accomplished in the second table in Figure 7-15, lower left. We then selected the horizontal lines of data that contained the three best porosity values and the three worst porosity values. The data was rearranged by descending porosity rating, as shown in the table to the upper right. That made it possible to continue the study and to observe that two variables—biscuit size and die temperature in degrees Fahrenheit—clearly separated the good and bad values.

Data Ranking Table Example

Sample Results Data for Illustration

Sample Number	Porosity Rating	Biscuit Size	Fast Shot Velocity IPS	Slow Shot Velocity IPS	Die Temp Degrees F	Water Flow GPM
1	4	3.00	34	10	450	18
2	1	1.75	38	12	425	21
3	4	3.25	41	11	380	19
4	2	1.75	43	9	420	20
5	5	3.25	37	10	385	18
6	3	2.50	41	8	400	20
7	4	2.50	39	9	370	18
8	5	3.00	45	12	360	20
9	1	1.25	34	10	460	22
10	2	2.25	34	11	430	20
17	1	1.50	44	10	440	21
18	0	1.50	35	8	425	22
13	3	2.75	38	9	390	20
14	5	2.75	36	11	400	19
15	2	1.50	40	9	420	21
16	1	1.25	38	11	440	21

Final Ranked Data Arrangement

	Porosity Rating	Biscuit Size	Fast Shot Velocity IPS	Slow Shot Velocity IPS	Die Temp Degrees F	Water Flow GPM	Sample Number
	0	1.50	35	8	425	22	18
Good	1	1.75	38	9	425	21	2
	1	1.25	34	10	460	22	9
	1	1.50	44	10	440	21	17
	1	1.25	38	11	440	21	16
	2	1.75	43	9	420	20	4
	2	2.25	34	11	430	20	10
	2	1.50	40	9	420	21	15
	3	2.50	41	8	400	20	6
	3	2.75	38	9	390	20	13
	4	3.00	34	10	450	18	1
	4	3.25	41	11	380	19	3
	4	2.50	39	9	370	18	7
	5	3.25	37	10	385	18	5
Bad	5	3.00	45	12	360	21	8
	5	2.75	36	11	400	19	14
Separation?		Yes	No	No	Yes	No	–

First Rearranged Data for Analysis

Porosity Rating	Biscuit Size	Fast Shot Velocity IPS	Slow Shot Velocity IPS	Die Temp Degrees F	Water Flow GPM	Sample Number
4	3.00	34	10	450	18	1
1	1.75	38	9	425	21	2
4	3.25	41	11	380	19	3
2	1.75	43	9	420	20	4
5	3.25	37	10	385	18	5
3	2.50	41	8	400	20	6
4	2.50	39	9	370	18	7
5	3.00	45	12	360	20	8
1	1.25	34	10	460	22	9
2	2.25	34	11	430	20	10
1	1.50	44	10	440	21	17
0	1.50	35	8	425	22	18
3	2.75	38	9	390	20	13
5	2.75	36	11	400	19	14
2	1.50	40	9	420	21	15
1	1.25	38	11	440	21	16

Three Goods and Bads chosen from the data above shown in color.

Findings:
1. There is clear separation in biscuit size.
2. There was no clear separation of data in fast shot velocity.
3. There was no clear separation of data in slow shot velocity.
4. There is clear separation in die temperatures.
5. There was no clear separation in water flow GMP.

Conclusion:
1. Biscuit size and die temperature are candidates for conducting a designed experiment to eliminate the defect.
2. Biscuit size and die temperature may have a direct and indirect effect on the defect.

Figure 7-15. Data Ranking Table Example Using Comparison

After experimentation, it was found that biscuit size was indeed a contributing factor in the development of porosity in the casting under study. The temperature variable was found to be not significant.

The table in the upper left of Figure 7-15 shows that 16 individual samples were observed for each of the five suspect variables. The data was recorded as generated by the process variables at random intervals.

The first rearranged data for analysis table is shown in the lower left. Six lines are highlighted to reflect the three highest and the three lowest porosity ratings. These were 1, 5, 5, 1, 0, and 5; they reflect the best and the worst quality manufactured during the period. (One is the best and five is the worst.)

The final ranked data arrangement at the upper right was rearranged according to the ranking order of the porosity rating. You can see that there is clear separation of the data regarding biscuit size and die temperature. We used this information to determine the two traits to be used in an experiment to identify the cause of the porosity problem.

Industrial Problem Solving Simplified | 119

Since only three of each group were compared to establish the variables for the experiment, the confidence of the results was suspect. The sheet shown in Figure 7-16 provides a more thorough analysis, which in turn provides more confidence when five of each group were compared. These analyses show a relationship in the comparability to a regression analysis, which follows the group comparison.

Ranking Sample Group Data for Comparison

Instruction:
1. Using the data, compare at least 5 Good vs. 5 Bad duo pairs.
2. Compare whether the measured value trends higher or lower for each compared duo.
3. Initiate a table as shown below and enter Indicators for the data trend for each duo.
4. Determine if all the indicators point the same for each of the 5 duos.
5. All the indicators with the same orientation indicate a direct effect of the variable.
6. Indicators not identically aligned indicate no direct effect of that variable.

	Porosity Rating	Biscuit Size	Fast Shot Velocity IPS	Slow Shot Velocity IPS	Die Temp Degrees F	Water Flow GPM	Sample Number
1 GOOD	0	1.50	35	8	425	22	18
2 GOOD	1	1.75	38	9	425	21	2
3 GOOD	1	1.25	34	10	460	22	9
4 GOOD	1	1.50	44	10	440	21	17
5 GOOD	1	1.25	38	11	440	21	16
	2	1.75	43	9	420	20	4
	2	2.25	34	11	430	20	10
	2	1.50	40	9	420	21	15
	3	2.50	41	8	400	20	6
	3	2.75	38	9	390	20	13
1 BAD	4	3.00	34	10	450	18	1
2 BAD	4	3.25	41	11	380	19	3
3 BAD	4	2.50	39	9	370	18	7
4 BAD	5	3.25	37	10	385	18	5
5 BAD	5	3.00	45	12	360	21	8
	5	2.75	36	11	400	19	14

Variable Effect	Duo # 1 GOOD/BAD	Duo # 2 GOOD/BAD	Duo # 3 GOOD/BAD	Duo # 4 GOOD/BAD	Duo # 5 GOOD/BAD	
Biscuit	<	<	<	<	<	Significant
Fast Shot	>	<	<	<	>	Not Significant
Slow Shot	<	<	=	<	=	Not Significant
Temp	<	>	>	>	>	Not Significant
Flow GPM	>	>	>	=	>	Not Significant

Finding: The only sample duo comparison that meets the requirements is biscuit size, as all of the indicators point in the same direction. Comparisons that have equal or mixed results are considered to be not significant by themselves.

Conclusion:
Biscuit size has a direct effect upon the defect being studied.
The other variables do not have a direct effect on the defect.
Interactions by the variables cannot be shown by this type of comparison.
Past studies indicated that interactions must not be ignored in problem solutions.
Use the variable with a direct effect for any designed experiment required.

Figure 7-16. Ranking Sample Group Data Using Five Good and Five Bad Samples

The following example uses a couple of tools to show the results. Note that the use of five good and five bad resulted in a more accurate clue generation. This was due to the fact that the comparison of duos indicated that only the biscuit size showed a significant difference between the good and bad porosity ratings.

Regression Analysis

For those who insist on some type of statistical analysis, and who have some knowledge of it, Figure 7-17 provides an example that verifies the five good versus five bad sample visual comparison shown earlier. This study uses a method provided within the Analysis ToolPak in Excel. The acceptable results that are required are shaded. They are as follows:

1. The adjusted R value must be > 0.70. (It is 0.8074.)
2. The significance F value must be < 0.001. (It is 1.38E-06.)
3. Each of the acceptable P values must be < 0.05. (They are < 0.05.)

Regression Analysis using Excel "Data Analysis"

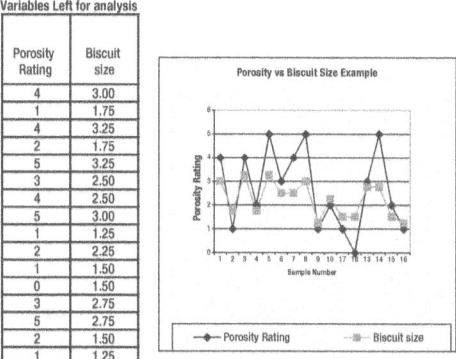
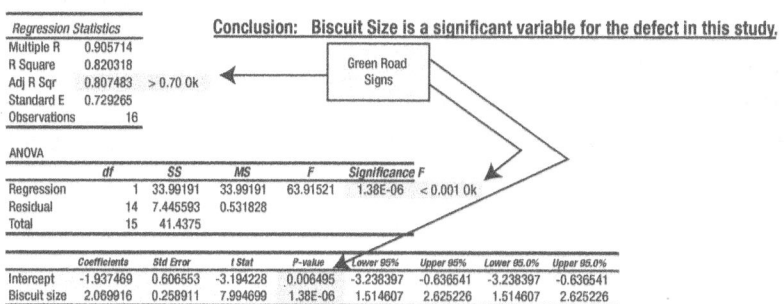

Figure 7-17. Regression Analysis Using Excel

The regression analysis example was calculated using Excel with the same information shown in Figure 7-16. The analysis shows that the biscuit size was the important factor in the defect being studied. This was the same result as in the group comparison study.

This does not mean that biscuit size is the determining factor for eliminating all types of porosity. It simply indicates that the porosity defined and evaluated in the ranking system was affected by the biscuit size. Verifying comparable test results using five of each sample or a formal regression analysis allows you to use either system.

Again, to maintain a high confidence in the result, compare only one trait at a time in each investigation. Because this was not a decision test or a validation test to prove certainty, we compared more than one characteristic in an attempt to determine clues about the important variables involved.

At this point, be aware that a comparison of two groups of five can also be used to confirm that a change was made. You can prove that materials, processes, repairs, or other conditions are effective by using a sum of extremes test and getting a result of 7 or greater.

As you can see, a sum of extremes test is a very useful tool and can be applied effectively for trial comparisons and to validate changes.

Fractional Analysis

This section deals with a manufacturing problem that required a fractional analysis to identify the causes of automatic assembly machine jams. This analysis is based on a matrix that records data generated in a system. The name is meant to relay that it is not a complete analysis of a system or its components. It differs in that it uses count data in a matrix to generate a numerical rating that can aid in the ranking of clues. I show examples that will help you understand. If you feel uncomfortable with the previous information, read Appendix A. It contains a complete explanation, with a sample problem and calculations. I include a problem and example later in this section.

Fractional analyses can help you decide which variables are important. They can designate variables to be used in designed experiments. They can also be used to generate clues when other insights are not present. (This is explained further in Appendix A.)

Consider this problem: Downtime was caused by jams at an engine block assembly line during a piston-stuffing operation. This occurred when an automatic feeder introduced pistons into the engine blocks. When the operation jammed, the entire production line became inoperative, because it was set up in a series orientation and the line could not move past the jammed loading station.

Industrial Problem Solving Simplified

The facility had one engine block assembly operation with two piston stuffers and had two separate machining lines supplying the engine blocks for assembly. The machined blocks were then transferred to the assembly line, where cases were assembled into automobile engines.

One of the assembly operations was to install pistons into the machined block bores at one of two piston-stuffing stations. These stations were in a series, on the same production line. Although each of these stuffers had the capability to stuff any of the four pistons, they each stuffed two different pistons. These operations were identified as 2050 and 2060, respectively. (They were building four-cylinder engines.)

One day there was an inordinate amount of jams at the stuffing station, which created line downtime. Manufacturing suspected that the supplied pistons were not to specification. This action required an immediate response, as the cost of the downtime was prohibitive. A suspect piston load was removed from the assembly line and replaced with another load from a previous batch, which was processed without incident. Shortly thereafter, the problem returned and was accompanied by a loud "snap" whenever the operation was placed into the manual or automatic mode.

A further examination resulted in the following information (the italicized words were all meaningful in helping to define the problem):

- Because of the spikes in occurrence, the condition was *irregular* and not continual.

- It appeared to be related to *part size* and not to component assembly.

- The condition was the result of an *event* in that something did not fit.

- These failures were not due to *part strength*.

- There was *work applied,* as the piston was squeezed to compress the rings prior to insertion into the bores.

- It was an *assembly* problem that involved *throughput* that allowed a *malfunction*.

- The failure occurred during *processing* and *assembly* on *both* stuffing operations.

- Because of the nature of the fault, it was not possible to determine if there were any witness marks (scratches or markings) in any area or region.

We decided to search for *unusual random patterns*, as the failures happened on both *shifts* on multiple *shipments* and the *design* had not been changed. All of these bullet points pointed to a *technical* problem with the assembly *block and virgin piston* components.

Fortunately there was no other plant affected, and recreating the failure with photographs was not possible. A random sample of the pistons and rings from the suppliers appeared to be within specifications. Even with this information, a summation of this information still did not allow a more accurate definition of the problem even though many variables were eliminated.

Normally at this point, at least five pistons and piston rings would have been captured and identified from each of the good and suspect piston loads. However, measurements were made to ensure that the parts were to specification. In addition, all individual data would have been subject to analysis to determine if there was clear separation of any of the characteristics measured. Since the piston assemblies were too complex to measure locally, five samples of both good lots or jamming pistons lots that had failed were returned via air shipment to the supplier for immediate observation and measurements.

We developed a piston-stuffing concept diagram and data sheet to define the defect characteristics, which were still unclear. These sheets indicated that there could be multiple sources of the problem, as shown on the concept diagram and data sheet in Figures 7-18 and 7-19.

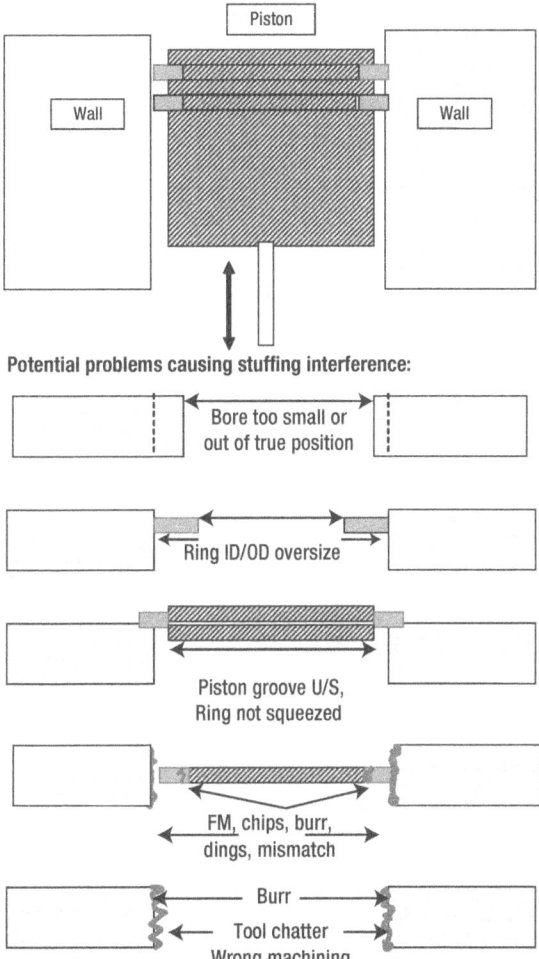

Figure 7-18. Piston-Stuffing Concept Diagram

Piston Stuffing Data Sheet (Revised)

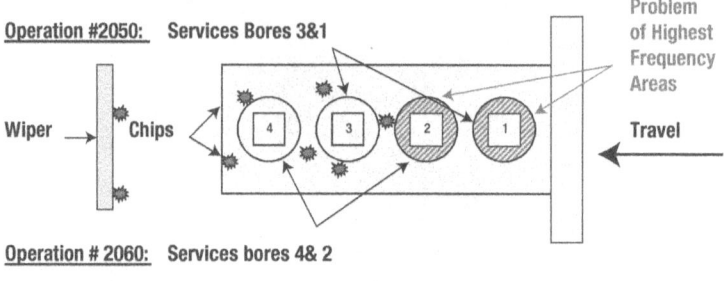

When Stuffing: Bores stuffed first are #1 then #3 on 2050, then #4 then #2 on 2060.

Total Stuff Pareto:		Op 2050 Stuff Pareto: Stuffed #1 then #3		Op 2060 Pareto: Stuffed #4 then #2		Stuff was Reversed	
#1	25	#1	19	#1	6	#1	5
#2	32	#2	1	#2	31	#2	0
#3	14	#3	13	#3	1	#3	0
#4	16	#4	1	#4	15	#4	0

Pareto

Findings:
1) Chips are not removed from case after machining at Op # 120 A or B.
2) Stuffed bores are excessive from machining A & B on different days.
3) Pareto of stuffing problems favors trailing bores #2 and #1.
4) #2 and #1 bores stuffed with east pusher on both stuffers.
5) When Op #2050 stuffs: #1 stuffs first & has more stuff problems than #3.
6) When Op #2060 stuffs: #4 stuffs first & has less stuff problems than #2.
7) First and last bores stuffed have the highest incidence of stuffing problems.
8) Wiper sweeps chips and FM into bores and may cause jam-ups.
9) Some pistons create a loud snap when manually inserted after a jam.

Figure 7-19. Piston-Stuffing Data Sheet

A revised data sheet—dissimilar from the one normally used on the line—was immediately created at the failing operations to capture as much information as possible (see Figures 7-19 and 7-20). These data included piston lot number, date of block machining, machining operation A or B, bore that jammed, and stuffing operation number (either 2050 or 2060). In addition, other observations were made of the troubled process to collect accurate current data relevant to the study. It is important to collect current data to allow accurate analysis of each problem. In most cases, historical data might have different time variables affecting the outcome, making it less useful.

Piston Stuff Interaction Table

	Jamming Frequency						
			PISTON Good		PISTON Suspect		
			Mach A	Mach B	Mach A	Mach B	
		Bore 1	2				
	Op 2050	3			1		
		2				1	
		4	1				
		Bore 1	3	1	2	1	
	Op 2060	3	1				Most Oper.
		2	6		6		Most Bore
		4	3		2		
			Most Frequent		Most Frequent		
	List of Jams:						
	1 Fail on #2 bore (B Machining) on Op 2050 with Suspect Pistons 2/16						
	2 Fail on #2 bore (A Machining) on Op 2060 with Good Pistons 2/15						
	3 Fail on #2 bore (A Machining) on Op 2060 with Suspect Pistons on 2/16						
	2 Fail on #2 bore (A Machining) on Op 2060 with Good Pistons 2/14						
	2 Fail on #4 bore (A Machining) on Op 2060 with Good Pistons 2/14						
	3 Fail on #2 bore (A Machining) on Op 2060 with Suspect Pistons 2/14						
	2 Fail on #4 bore (A Machining) on Op 2060 with Suspect Pistons 2/14						
	2 Fail on #1 Bore (A Machining) on Op 2050 with Good Pistons 2/14						
	1 Fail on #4 Bore (A Machining) on Op 2050 with Good Pistons 2/14						
	1 Fail on #4 Bore (A Machining) on Op 2060 with Good Pistons 2/14						
	1 Fail on #3 Bore (A Machining) on Op 2060 with Good Pistons 2/14						
	2 Fail on #1 Bore (A Machining) on Op 2060 with Suspect Pistons 2/14						
	1 Fail on #1 Bore (B Machining) on Op 2060 with Suspect Pistons 2/14						
	1 Fail on #2 Bore (A Machining) on Op 2050 with Suspect Pistons 2/14						
	3 Fail on #1 Bore (A Machining) on Op 2060 with Good Pistons 2/14						
	1 Fail on #1 Bore (B Machining) on Op 2060 with Good Pistons 2/14						
	1 Fail on #2 Bore (A Machining) on Op 2060 with Good Pistons 2/14						
	Conclusions based on the sample:						
	1) Biggest cause of variation is machining A at rating of 13.5						
	2) 2nd largest cause of variation in Op 2060 rated 12.5						
	3) 3rd largest cause is machine/stuffer interaction rated 9						
	See explanation in **Appendix X** for method of determination						

Figure 7-20. Piston-Stuffing Interaction Table

The collected data is presented in the piston-stuff interaction table (see Figure 7-20). It is laid out in an arrangement with four variables:

- Machining performed at operation A or B
- Identified good or suspect piston assemblies used

- Stuffing operation 2050 or 2060
- Individual bores in the block that were loaded

The defect summarized in each block is the number of failures associated with each of the variables for that specific block in the matrix.

The conclusions in Figure 7-20 summarize the fractional calculations, which are explained and shown on the following pages. If you are uncomfortable with accepting the analysis results that will be given without first knowing the method used, refer to Appendix A, where the process of fractional analysis is explained.

These conclusions indicated that machining A contributed the most to the piston-stuffing problem, as it had the highest calculated effect rating of 13.5. Even though all of the machining characteristics were found to be within specification, it was the largest contributor. Operation 2060 had the second highest effect rating of 12.5, which indicated that it was much more prone to failures than was operation 2050. In addition, the calculations show that there was an interaction with effect ratisng of 9.0 between machining A and operation 2060, especially on bore 2, which had the highest incidence of failure.

There were other incidental effects, which were not as predominant as those previously listed. These were:

- Bore-to-bore variation had an effect rating of 3.5
- Piston quality and machining variation interaction had a rating of 3.0
- Piston quality variation had the lowest effect rating of 2.0

Conclusions Based on Current Samples[3]

The conclusions in this section are the result of calculations that will be presented in the examples. The variable listed and the results that follow indicate the relative effect that the variable has on the process being evaluated. That is, the greater the number value shown the greater the variable's influence in causing jams at the loading stations.

1. Machining A caused the most failures, with an effect of 13.5. with 27 jam-ups.
2. Operation 2060 caused the second highest variation, with an effect of 12.5 with 25 jam-ups.

---

[3]Refer to Appendix A for more comprehensive calculations and discussion.

3. Machining A and operation 2060 have an interaction, with an effect of 9.0. (See example for interaction jam-ups.)
4. The direct, indirect, and interactions are equally strong with operation 2060.
5. Something is wrong at the operation 2060 stuffing station.
6. Machining effects are larger than operation-stuffing effects.

Recommendation: Machining A must be refurbished and retargeted to nominal.

What had appeared to be a piston-quality problem was in reality a machining setup and stuffer operation problem. Retargeting machining operation A to be more like operation B and adjusting the alignment and operation of stuffer 2060 prevented recurrence of the problem.

There were other variables that were acting during the study, but they were not as significant as those that caused the majority of the stuffing problem. The load of pistons also contributed, but was not as responsible for the jams as were the other factors. Remember, use only current data in evaluations as historical data may contain other variables that can confound the investigation.

Now that the overall scope of the project has been presented, let's look at the method used to determine the association of variables and the assignment of their responsibility to the problem. (See Figure 7-21.)

Piston Stuffing Fractional

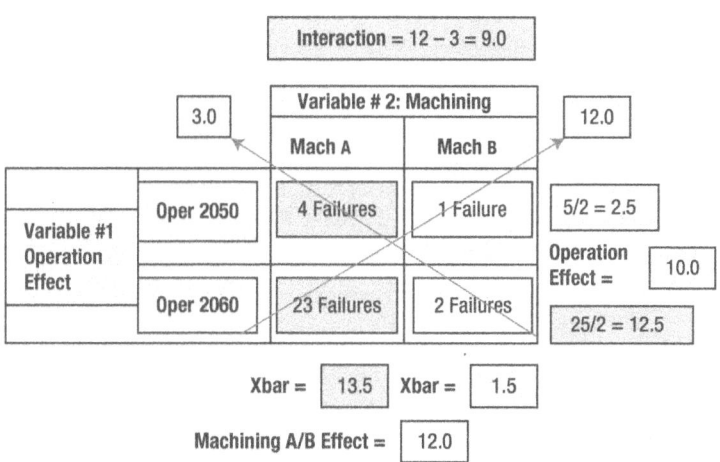

Figure 7-21. Piston Stuffing Fractional Example

Chapter 7 | Choose and Use Analysis Tools

The steps listed next were used to set up the matrix. The main horizontal variable was addressed as machining, whereas the vertical variable was addressed as the stuffing operation. There were two machining operations designated as A and B. There were two stuffing operations designated as operation 2050 and 2060. The data in each cell of the matrix was determined by adding the number of jams that occurred with the conditions that correspond to the squares in the matrix in Figure 7-20. Operation 2050 had four failures with blocks machined on line A, whereas it had 1 jam with blocks machined on B. Operation 2060 had 23 jams with machining A while only two with machining B. This resulted in the following first four assignments listed. The numbers generated on the display (Figure 7-21) are shown in steps 5 thru 13.

1. The matrix contains two main variables—operation and machining.
2. Operation has two levels (2050 ad 2060).
3. Machining has two levels (machining A and machining B).
4. The number of failures is placed in each box for operation and machining pairs. (Machining A had four failures in operation 2050, whereas machining B had two failures in operation 2060.)
5. Operation 2050 had five failures with Machining A plus B. The average calculation for this severity is (4+1)/2 = 2.5.
6. Operation 2060 had 25 failures with machining A plus B. The average calculation for this severity is (23+2)/2 = 12.5.
7. Operation effect = 12.5 − 2.5 = 10.
8. Machining A had an average of (4+23)/2 = 13.5.
9. Machining B had an average of (1+2)/2 = 1.5.
10. Machining effect = 13.5 − 1.5 = 12.0.
11. Machining A, Operation 2050 vs Machining B, Operation 2060 = (4+2)/2 = 3.0.
12. Machining A, Operation 2060 vs Machining B, Operation 2050 = (23+1)/2 = 12.0.
13. Interaction between operations and machines is (12 − 3) = 9.0.

Machining accomplished with process A at 13.5 has the largest effect on producing jams when combined with piston-stuffing operation 2060. A high interaction is also present (9.0) when they interact together.

Tests for Clue Generation or Verification

You can perform clue generation or verification by comparing two populations. The different populations may be referred to as the "best" and the "worst," or can have any designation. It could be the comparison of two different materials, machines, tools, processes, methods, suppliers, or environmental conditions that are proposed or on hand.

In the following test, the desired result is zero. A water filter manufacturer wanted to change the plastic cover for its pitcher because of customer complaints that the covers cracked when dropped to the floor. A replacement plastic was proposed, but there was uncertainty as to whether the covers were better with the proposed new material.

The first order of business was to devise a rating system. 0 equaled no cracks whereas 5 signified broken in half. The second step was to devise a drop test from a 40-inch height to replicate the drop from a countertop to the floor. The third step was to randomly drop and rate the damage of three of each of the old plastic covers (the dotted columns in the chart) and three of the newer plastic covers (solid columns in the chart). The results were depicted in graphic form (see Figure 7-22).

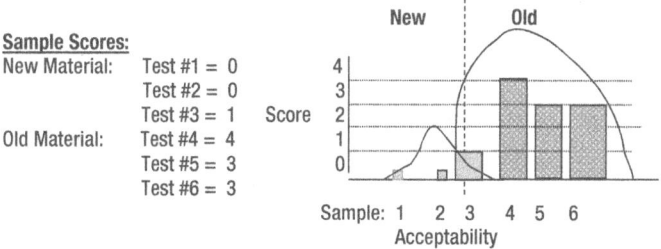

Figure 7-22. Old Material Versus New Material Test

As you can see, there was adequate separation of the data. The new covers cracked less. Although the new material is better than the old, it was still subject to cracking. One of the new covers cracked slightly, but three of the old covers cracked severely. Unless an even better material is tested, the new material should replace the old.

New plastic covers still break occasionally but appear to be stronger than old ones. The customer requested the newer replacement material covers until a more suitable stronger cover could be found.

Figure 7-23 shows another example of a comparison test. The test was conducted to determine clues as to what was causing the gears to bend in the gearboxes of juvenile electrical-powered vehicles. We disassembled three failed and three operable gearboxes and measured each component. The following characteristics were suspect: Hub Gear ID, Gear Post Diameter, Third Cluster Bore Pin, Pin Diameter, and Third Cluster Bore Gear. Each of the measurements for the failed and acceptable parts had populations that overlapped with the first four characteristics. The data did not overlap for the third cluster bore gear when each sample was measured in four different locations.

3rd Cluster Bore Gear

Failed Units			Good Units		
#1	#2	#3	#4	#5	#6
0.382	0.384	0.383	0.379	0.380	0.379
0.383	0.384	0.385	0.381	0.380	0.380
0.384	0.384	0.385	0.379	0.379	0.380
0.383	0.385	0.384	0.380	0.380	0.378

Range 0.382-0.385 inches = Tighter Fit   Range 0.378 to 0.380 inches = Looser Fit

Figure 7-23. Third-Cluster Bore Gear Data Table

There is clear separation of the data, as shown in the two halves of the table in Figure 7-23. The difference of only a few thousandths of an inch created an interference fit that resulted in premature toy failure. The difference was with one of two gears in the mold used to manufacture the gears.

Besides the clear separation in the data, there was 95% confidence that the good and failed gear units came from different populations. However, in situations like this, use caution. In this case, the spacing between the populations was small and failures were still possible with the perceived good gears if a larger outlier[4] was created.

This type of test can be used to verify that corrective action placed into effect is effective. There is good confidence if the change data is clearly better and statistically separated from the before (unacceptable) data values. Always confirm the effectiveness of changes after they are made.

---

[4]An outlier is a point of data that falls far away from the mean of the distribution from which it came. It is an anomaly in that it is significantly distant from the rest of the samples from which it came. In this case, the outlier could mingle with a sample from the other part distribution if they were not sufficiently far from each other.

Good versus Bad Comparisons

Sometimes it is necessary to determine the nature of the differences between acceptable and unacceptable products. In this case, it is sometimes possible to use the comparisons method described previously. This is an example of testing that you can use with the application of test achievement comparisons.

Comparison of Good and Bad Assemblies

A juvenile product manufacturer designed a door gate to restrict access of toddlers to a specific room. The gate was designed to be installed within the door frame and was actuated by stepping on a locking lever. Actuation of the lever forced the walls of the gate to push against the door frame, ensuring the doorway was sealed. This type of gate could be used at the second-floor level to prevent toddlers from tumbling down an unguarded stairway. It had to be a safe and reliable barrier.

Unfortunately, some test gates were not stable and did not provide a satisfactory barricade. A test was developed to verify that the gate would prevent a child from disassembling it from a doorway even if the child threw a tantrum and shook the gate violently. We attached a cylinder to an installed gate and applied a force of 40 pounds in repetitive cycles to a forward and backward rocking motion.

Some gates were highly satisfactory, whereas others fell out of the doorway with only a limited number of cycles. Characteristics of the three best and the three worst gates were compared, as shown in Figure 7-24. Only five of the items checked are shown in the illustration. The one that shows separation is most critical.

Gate	Cycles to Failure	Horizontal Tube	Vertical Tube	Pin Dist	Dim B
B1	360	1.011	1.002	0.015	1.288
B2	800	1.010	1.000	0.000	1.287
B3	1360	1.010	1.001	0.010	1.286
G1	11,880	1.014	1.008	0.011	1.289
G2	12,470	1.011	1.009	0.013	1.284
G3	12,500	1.014	1.012	0.019	1.288
	Overlap	Overlap	Separation	Overlap	Overlap

Figure 7-24. Toddler Door Gate Material Overlap

Components from three good toddler door gates (G1, G2, and G3) were measured and compared to components from three bad toddler door gates (B1, B2, and B3), which failed to stay in place. Some of the resulting ranked measurements were as follows:

The top three sections represent the identity and the measurements of vertical tube width in inches for the gates that failed to stay in place.

The lower three sections represent those good gates and the vertical tube measurements that exceeded the test bogey of 10,000 cycles of vibration that were deemed to be acceptable.

The only dimension that showed clear separation was the vertical tube width, with 0.006 of an inch between the separated values (see Figure 7-25). We discovered that the supplier used a different tubing source without prior approval. The supplier did not provide the certified material that was required to produce a safe operating gate. An unauthorized change therefore created failures.

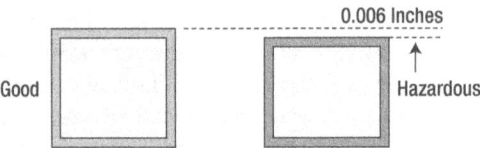

Figure 7-25. Drawing of Steel Tubing Supplied

This had to be controlled to ensure that the door gates would operate properly in the field. The supplier made the change and provided two different batches of tubing for the trial production run without authorization. It was necessary to establish a certification requirement for these materials for this supplier to prevent a serious injury or death.

By now you can recognize the significance of this method of comparison. It can generate powerful clues quickly. However, this system is flawed in that it is not completely reliable for decision making. Sometimes the variable that shows the separation difference does not affect the outcome that you are trying to evaluate. It is best to employ the comparison of five good and five bad samples so that you can observe any characteristic differences in the parts.

In addition, it is best to always strive for a minimum separation between the two populations of more than one sixth of the *widest* population. Distance is represented by the X shown in Figure 7-26. The distribution on the left is wider than the one on the right. If you divide it into six equal sections, you end up with a width as shown by X. The two distributions should be more than X distance apart to give you confidence that they are entirely different populations and that there should be no outlier data point that is well outside of its relative distribution.

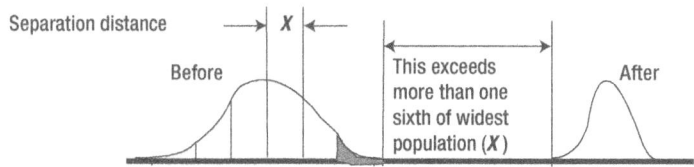

Figure 7-26. Required Sample Separation

The data spread from the "before" sample—when the bad items showed up—must be separated from the data spread of the "after" results—when you found replacement or known "good" components. If they overlap or touch, then you cannot have complete confidence that the results are different and significant.

Summary

This chapter discussed innovative tools and methods that you can use to investigate more complicated problems. They do require a more concentrated effort because actual measurements and data comparisons are involved. They will help you to move more directly to what may be causing the problem under study. The next chapter involves the use of some other tools that utilize visual inspections, flow charts, lot identification, sounds, and crack and broken component considerations.

CHAPTER

8

Use Innovative Analysis Tools

Step Seven

This chapter deals with the use of visual aids, including flowcharts, lot identification, and the evaluation of sound and broken components. It includes examples of failures and the solutions that were effective in correcting them.

The following tools enable the problem solver to focus on more easily solved problems that might not require in-depth evaluations. The innovative use of analysis tools is the seventh step of problem solving. When nothing else appears to work, these elements can allow your creativity to flourish.

Visual Observation

The most important "analysis" tool is simply to insist that everyone involved observe the problematic process or component in question. It is not uncommon for groups to convene in a conference room or office and intensely debate the cause of a problem without having observed the existing conditions. You can't solve problems effectively while seated in a conference room.

The following two examples lend credence to this point.

I was involved in a situation in which incorrect valve stems (see Figure 8-1) were added to the assembly line because samples were not provided or discussed. We didn't discover the difference between the two components until well into the manufacturing cycle. This is an example of why you need a visual understanding of each problem. It's so much easier to understand a verbal description when you've actually seen the faulty part or system. At the very least, you should provide samples of the part being discussed for examination. It should be mandatory that evaluative personnel and teams visit the site of the problem to understand the conditions.

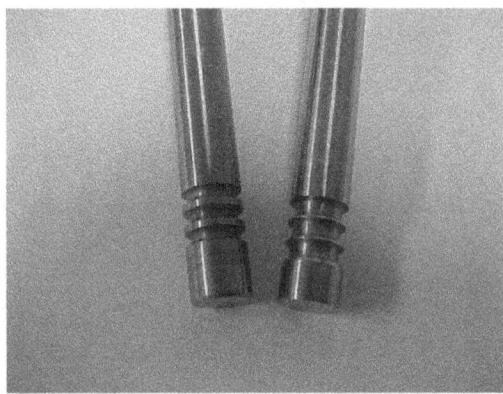

Figure 8-1. Photograph of Valve Stems; Note the Difference Between the Two

This problem with the valve stems was due to poor communications and training; a new untrained employee was helping to service the line. To prevent similar defective assemblies, we provided training and photographs of the components to the line.

Here's another example: When assembling engines, one must submit them to a torque test after the pistons are inserted to ensure that they conform to the requirements. At some point, there was a dramatic increase in torque failures. The line stopped, because the repair station was overloaded with line rejects. The failure not only created an excess amount of engines that had to be repaired, but it also generated a substantial loss due to the manufacturing facility's inability to sustain production.

Confusion takes over with any line stoppage. Parts are immediately taken to the metrology lab for measurement and analysis. Sample defective products are disassembled for evaluation. People capture and sequester product that is not to specifications. Attempts are made to restart the line and to capture data as the product is being manufactured. After a day or two, we discovered little slivers of foreign material underneath the assembly conveyor, but the origin of the material could not be determined.

In this case, we gathered those involved for a walk along the assembly line to observe each operation. During that walk-through, someone noticed that a change in piston rod design was allowing a piece of flashing attached to the piston to fall into the engine after the piston was installed. We discovered this clue by observing a significant accumulation of minute sliver-looking components under the piston-setting operation machinery.

Since the small chips and slivers were not present before, we determined that the new piston design or machining operation was contributing to the problem. We found that the slivers should have been addressed at the supplier location and were to be removed before they were shipped. A step in the supplier's process was not in place when we received the first shipments. This contributed to torque test failures with the new components. The biggest clue was that the slivers were under the assembly machinery, which we noted by visual inspection. The arrows in the photograph show where the chips and slivers originated (see Figure 8-2). The old style (on the left) had square shoulders, whereas the new design (on the right) had sloped shoulders, which is where the slivers originated.

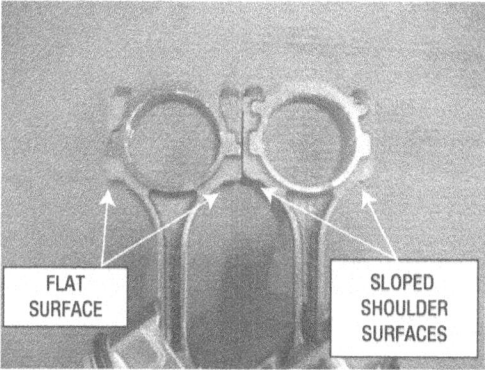

Figure 8-2. Photograph of Piston Shoulder Differences

In summary, a visual observation of the process can be very rewarding. It is a simple tool that is greatly underused by many manufacturers experiencing problems.

Tip When problems arise, someone should always be assigned to view the current process and to capture samples of the items leading to the crisis.

Unfortunately, walk-throughs are not done in most instances. In my experience, whenever a major problem is found, there's an immediate call for a meeting. Most people attempt to define the current problem in relation to their last experiences, without observing the present, applicable conditions. Nonetheless, everyone involved must be visually aware of the most recent conditions before adjourning to the conference room for problem-resolution discussions.

Figure 8-3 shows the minuteness of the sliver that was causing the torque failure problem when it was inadvertently deposited into the engine.

Chapter 8 | Use Innovative Analysis Tools

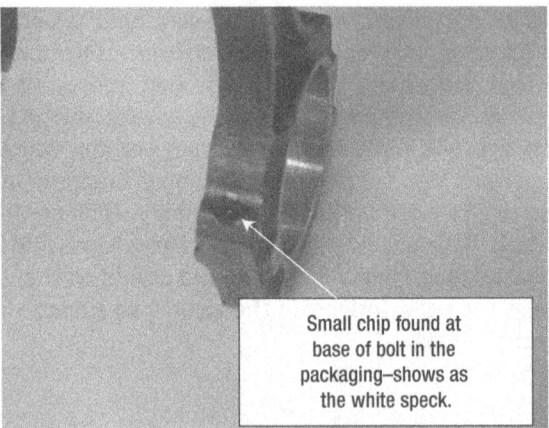

Figure 8-3. Photograph of Foreign Matter Found on the Piston

When necessary, you should plan to call your affected customers immediately to alert them of any difficulty that may affect the operation or use of the product. In this case, an unqualified design change allowed a sliver of machining particle to cling to the new components until the installation operation. At that point, a minor vibration shook them loose.

Lot Identification Flowcharts

Flowcharts can be a valuable way to view how specific components are routed. They can be used to verify that the process has been designed correctly to ensure proper component flow.

In addition, they can be used in the event of quality-assurance problems that require production stops or recalls. The example I discuss shows how you can identify a recalled product, while restricting the amount of questionable product that's currently in stock or in the customers' possession. Doing this successfully requires planning and identification.

However, before you learn about using flowcharts, it is important to note that flowcharts depend on the correct actions of the personnel involved. Control plans must be established to ensure that quality requirements and certifications continually adhere to and comply with approved process methods. Consistent work must be unfailing or you can expect problems.

It is not uncommon to have components wander from an acceptable flow path or to become mixed with unauthorized components by accident. Consider these scenarios:

- A part removed from an assembly line because it requires rework isn't introduced back into the assembly line at the position where it was removed.

- Parts not properly identified to begin with are removed from a process only to be introduced at a later station without the required electronic checks being performed.

- Defective or incorrect parts are reintroduced into the system because someone picked them off of the floor.

- Parts become mixed because they had not been properly tagged after a first-piece inspection.

It is important to walk the process to ensure that parts are properly handled during every step.

Whenever parts are removed from the specified routing route, they should be marked, tagged, or otherwise identified. This will help prevent them from being improperly introduced somewhere else in the system. Containers, and the parts inside them, should be identified in the same manner. You should ensure that identification is removed from the containers after they are emptied. Scrap items placed into containers should be identified and physically destroyed to prevent reuse in the production process.

Figure 8-4 shows a simplified example of the flowchart that traces the movement of parts through an assembly process. It is not the complete process, but it will give you an understanding of the concept. Design and process failure mode effect analysis, as well as control plans and job instruction preplanning, are all necessary in order to establish the routing and identification processes.

Chapter 8 | Use Innovative Analysis Tools

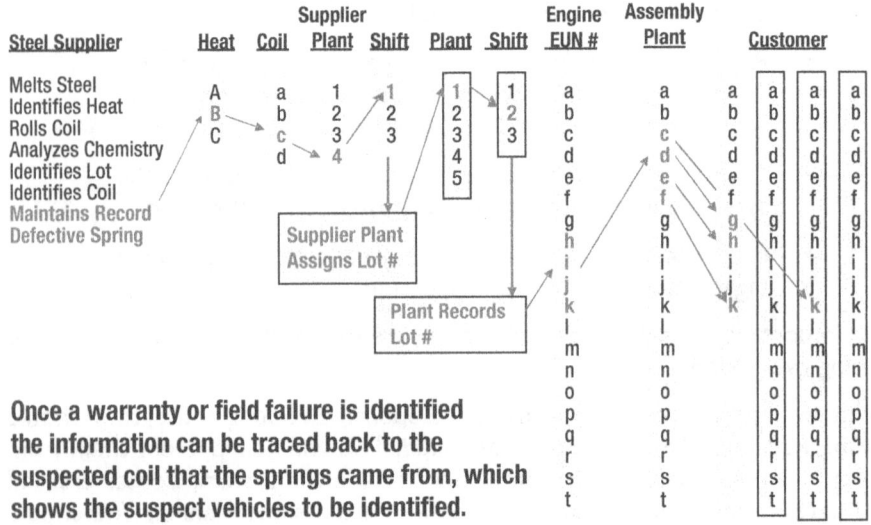

Figure 8-4. Lot Identification Flowchart

The chart shown in Figure 8-4, hypothetical and condensed, attempts to reflect the path of spring coils that will be used at an assembly plant and then be shipped to customers. The steel supplier maintains records of the heat, the metal consistency, the coil (identification of the roll produced), the plant it was produced in, and the shift on which it was produced as a minimum identification system. Then it is shipped to the manufacturing customer, who creates springs out of the rolled coils. This manufacturer also keeps track of product identifiers by recording the plant, shift, relative engine process identifications, and other important criteria that identify the process used at that plant. The engine is sent to one of many assembly plants, where it is installed into a vehicle and sent somewhere via the purchaser network, from which individuals can then purchase the vehicles.

Now suppose that after a length of time the original steel supplier realizes that a mistake had been made in their process, and that a certain batch of coils was defective and could result in a safety hazard. The supplier had the foresight to record the identifiers and the flowchart that it maintained.

Industrial Problem Solving Simplified

They identified not only the heat and the coil but also the customer to whom they had shipped the defective coil.

Notifying the engine customer alerted them to the problem, which allowed them to trace the coil through their manufacturing system and to trace it to the assembly plant to which it had been shipped.

The assembly plant, by maintaining its records well, could determine which dealer they shipped the engine to. The assembly plant and dealer would be aware of the customer to whom the vehicle had been sold. This customer would receive a recall notice based on the information captured in a flow plan.

This type of plan is not dissimilar from a flow plan that follows raw material from entry into a manufacturing plant through each operation and inspection. It can be used to ensure that parts removed from the designated flow path be reintroduced into the flow path at acceptable designated stations. The lack of a designated flow path can result in serious process problems of scrap or rework.

Unfortunately, all problems are not identified or sequestered before they affect the customer. The importance of being able to identify and trace a lot or part is extremely important in containing recall costs due to quality spills. Hence, the importance of the flowchart. Since the process flowchart can track the components from the cradle to the grave, the flowchart makes it possible to backtrack from the point of failure to the point of origin. By verifying the identification information on the defective part, recalls can be issued only to affected customers, and not to the greater customer bank.

Tip Since you can use identification numbers with pattern serials, processes, components, vehicles, and customers, they can also be used to evaluate problems. Be sure to use all of the information available when you're collecting data and to follow the identifiers back to the source of the problem.

In addition, always attempt to identify components that pass through similar operations on any multipath system. You can do this by applying a different colored paint dot or mark. This will be of enormous value when you analyze future problems. Remember the bent connector problem that was affecting the truck assembly plant discussed in Chapter 3? Only one of the test machines was causing the problem. The plant automatically recorded which engine was being tested on which machine, and that data trail made it much easier to resolve the problem. The flow path in that case involved checking the parts traveling down the assembly line; each operation and deviation from a common path was recorded. The engine identification numbers indicated

Chapter 8 | Use Innovative Analysis Tools

that all the damaged connectors were from one test stand, so it was clear that we needed to inspect and repair that test stand, which was identified in the flow path.

The value of using visual clues cannot be overstressed. They can provide a quick analysis of the problem at hand. Consider the next three examples; each involved visual clues that led to a quick response to the problem.

Case 1: A Malformed Bearing

The first case involved malformed bearing surfaces. Upon inspection, we observed a crease at the base that wasn't present on previous production components or samples (see Figure 8-5).

Figure 8-5. Malformed Bearing Surfaces

Also, the formed-tab length was extended on the bad part with the oil groove on the left side than on the good part without the oil groove (see Figure 8-6). The elongated tab, which stuck out too far, caused an interference at its base when it was attached to the mating surface area.

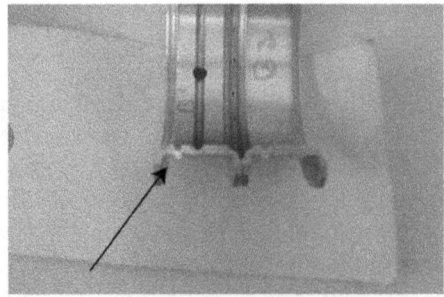

Figure 8-6. An Elongated Tab on the Bearing Surface Interferes with the Part It Mates To

Merely comparing a sample of a current troublesome part and a previously accepted production run component can have beneficial effects if there is a visible difference. With practice, it takes only a few moments to identify any differences.

In this case, the bearing supplier was notified about the discrepancy causing the problem, and they undertook actions to correct their forming process. The most current load of defective bearings could not be repaired and were subsequently scrapped. We requested a new shipment, which contained acceptable bearing lots that were employed in the assemblies. We took no further actions at our facility, but the supplier was instructed to certify all future shipments to ensure that the bearings were acceptable.

Case 2: Parts Missing

The second example involved a rash of complaints that an assembly line was receiving too many subassembly components with a required part missing. The failure to utilize first-time quality parts required us to complete costly rework operations.

Upon investigation, it was found that an O-ring was missing when the parts were received from the supplier. The supplier could not understand the complaint because they said the components went through a 100% inspection before they were packed for shipment.

We conducted a visual review of the subassemblies that had been shipped and received at the assembly facility. In the bottom of the containers, we found O-rings that had broken and become disassembled from the handle of the oil level indicator. It was inconceivable that a rubber O-ring, which resembled a thick rubber band, could disassemble itself or corrode and fail during a shipment to the place of assembly. So what was going on? We needed to investigate further.

Figure 8-7 shows that the O-rings could exceed a 200% elongation test as detailed in the specifications. This indicated that the failure was not due to excessive stretching applied during installation by the assemblers. The O-rings themselves were of sufficient form when observed; the diameters and thicknesses of the failed and assembled components were the same. Some O-rings remained on the oil level indicator handles and functioned as designed. There appeared to be no differences between the failed and assembled O-rings that would indicate that they were defective. Their plasticity, inside diameter, and thickness were to specification.

Chapter 8 | Use Innovative Analysis Tools

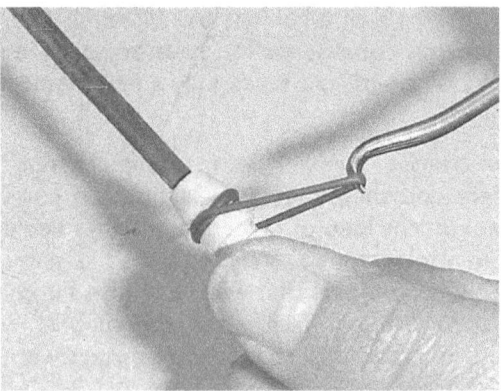

Figure 8-7. Inspection Showed that the O-ring Retained Its Elasticity as Received

The sharp fin on the indicator caused by the mismatch of its two halves was much more significant. The sharp edge (fin) actually caused the O-ring to slice apart. So the O-ring contribution to the failures was eliminated.

What we found, through observation, was a difference in the matching of the plastic oil level indicator handles on which the O-rings were installed. Figure 8-8 shows the desired and defective conditions where the two handle parts meet for matching during the plastic molding operation. This defective condition is generally caused by missing or severely worn guide pins and bushings on the pattern equipment. Since the patterns did not match as they should, shifting of one half of the mold caused the mold mismatch. This mismatch created a sharp corner on the handle, which formed a stress point for the O-ring. That stress point caused the separation.

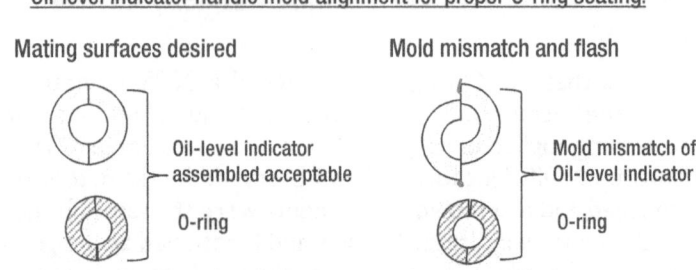

Figure 8-8. Oil-Level Indicator Handle Mold Alignment for Proper O-ring Seating

Looking at the components helped us quickly resolve the difficulty. We sequestered and inspected all of the oil level indicators at the supplier, those in shipment, and those at the customer, which prevented the use of any more unacceptable parts. The supplier immediately inspected their process bushings and pins and replaced the questionable items. They also initiated process checks, which had not previously been conducted, to prevent the condition from recurring.

Case 3: Putting Excess Iron to Good Use

This third example is a visual one made with the mind's eye. Let me explain.

One metal casting plant is capable of pouring four different types of iron. It pours piston iron, cylinder iron, high-carbon-equivalent iron, and nodular iron. Piston and cylinder iron generally differ only in the silicon content. High-carbon-equivalent iron has a higher carbon content than the piston and cylinder iron. The nodular iron has carbon contained in small nodules as compared to flakes, which are present in the other three types of iron. These different iron types are all necessary because different parts require the different properties of the iron being used.

All of these mixes are produced in batches on the same production line and are all mixed in the same furnace that pours the iron into the different molds. When one part is finished, the excess iron is poured into pig boxes in the basement, where it is allowed to solidify to be melted again on another day. Some of the pigs formed unusual shapes when they overflowed their boxes and were allowed to solidify on the basement floor. There were tons and tons of iron being melted, solidified, and re-melted each week. The cost of these reprocessing changes exceeded $50,000 per year.

One day, while observing the operation, a young engineer saw what he believed to be a barbell forming in the overflowed iron in the basement. He had an idea and enlisted the aid of senior managers to investigate the use and sale of the recycled iron to eliminate melting costs and rehandling operations.

His efforts paid off: A manufacturer of weights and other gym equipment bought the excess iron in the form of barbell weights at a significantly reduced price than what they had been paying at another foundry. After all, a weight lifter doesn't care about the type of iron that they are lifting, they are concerned only with the adequate weight and repetition of their exercise.

Had no one looked at the operation and seen the possibility, the waste would probably continue to this day. With that in mind, you can move to the next section, which deals with lot identification.

Sound Problems

Sound problems are present in all types of applications. They can be present in manufacturing operations during the machining, assembly, installation, and shipment of products. Customers may complain about strange noises that accompany poorly functioning parts. In general, they do not lend themselves to easy solutions. Noise exhibits itself in the form of chatters, grinding, snaps, rattles, chirps, gurgle, squeaks, squeals, hisses, screeches, whines, and whistling patterns, to name a few. Noise reverberates, which makes it difficult to determine where it's really coming from. It can bounce off of reflecting surfaces or obstructions or be diminished by absorbent materials.

This section attempts to help you develop sensitivity to noise and sounds that accompany industrial problems.

You must evaluate and define noise problems specifically to facilitate elimination. Figure 8-9 shows some relationships between the sounds experienced and the suspected causes.

Sound:	Potential Cause:	Sound:	Potential Cause:
Chatter	out of round	Roar	excessive escape rate
Chirping	bad alignment	Scraping	interference
Clank	metal on metal	Screeching	loose parts
Clicking	contraction	Shrieking	angry customer (Sorry!)
Clicking	tension release	Slap	breaking / loose belt
Crunch	material fracturing	Snapping	discontinuous contact
Grinding	surface interference	Squeaking	bad alignment
Growling	mismatch / Interference	Squeal	loose belt
Gurgle	entrained gas	Squeal	out of alignment
Hiss	pressure release	Swoosh	rapid fluid movement
Rattling	loose hard parts	Whining	uneven surfaces

Figure 8-9. Assembly Noise Study Table Chart

Telltale sounds can help identify the cause of a problem. If the sound reverberates in an enclosed area from a direction that could not possibly generate it, look in a direction exactly opposite from the perceived direction. It might be that the sound is reflecting off of a sympathetic surface.

You also want to observe any vibrations or fluttering associated with the noise. Identifying any contributing causes allows you to eliminate unrelated variables, which in turn helps you focus in on the source.

Industrial Problem Solving Simplified

Noise Problem Matrix

Sometimes you'll need to disassemble and reassemble units and test for differences in sound under different conditions with different components. Figure 8-10 shows one method for comparison. You measure the noise differences when different components are combined, such as switching noisy engine block balancer components with quiet and acceptable block components. This type of matrix can provide useful information in noise problem studies.

Balancer Assembly Noise Study: Measured Noise Rating

	RPM	Good Block & Bad BSA RPM vs RPM	Good Block & Good BSA RPM vs RPM	Bad Block & Good BSA RPM vs RPM	Bad Block & Bad BSA RPM vs RPM	Bad BSA to Bad BSA Bad Block	Bad BSA to Bad BSA Good Block	Good BSA to Good BSA Bad Block	Good BSA to Good BSA Good Block
	1400	70	50	30	20	70	20	30	50
	2600	600	40	300	50	600	50	300	40
Difference		530	10	270	30	530	70	270	10
		3rd				2nd			

Sector Difference	RPM	Bad Block to Good Block - Bad BSA	Bad Block to Good Block - Good BSA	Bad BSA To Good BSA With Bad Block	Good BSA With Good Block
	1400	50	30	40	30
	2600	550	20	300	10
		1st			4th
		5th			6th

Findings:
1st Block to block differences are the most important to be evaluated (Largest Variation = 600 - 50 = 550)
2nd RPM differences contribute to 2nd largest family of variation (Variation = 530)
3rd Bad block and bad BSA combination assemblies are influenced by RPM (Variation = 530)
4th Differences between good / bad BSA do not significantly affect the variation at different RPM in good blocks
5th RPM is part of the interaction as largest family of variation (1st) is affected significantly at the different RPMs.
6th RPM interaction is not present or minimal with good blocks.

Conclusion: Must find clear data separation in characteristic that interfaces with BSA in good / bad blocks.

Figure 8-10. Balancer Assembly Noise Study Matrix

This type of action is necessary when detrimental noise problems are present. In this case, we needed to study vehicle engines and their assembly components because it appeared that the combination of components used for assembly affected the amount of noise generated.

We conducted the study in order to determine which conditions eliminated the noise. The components we used for the analysis were two blocks that differed in the amount of noise that was produced at each of two RPM levels. The assemblies that were less noisy were considered to contain less noisy balancers and less noisy blocks. The engine assembly components that were noisier were specified as being bad balancers and bad blocks. (The difference in blocks was due to the machining, whereas the difference in balancers was due to their construction. I will not define these conditions and characteristics here.) The test was conducted at two different levels—1,400 and 2,600 RPM.

Chapter 8 | Use Innovative Analysis Tools

As Figure 8-10 shows, we conducted each of the following tests for the different component assemblies:

- Good blocks with bad balancers (BSAs)
- Good blocks with good balancers
- Bad blocks with good balancers
- Bad blocks with bad balancers

In addition, we compared each of the bad balancers to the other bad balancers, first on a bad block and then on good block. Similarly, we compared each good balancer to the other good balancer, first on the bad block and then on a good block. We took readings with a calibrated and available instrument and recorded the data into the table.

The lower-left part of the table shows the relevant data we calculated from the readings listed in the upper half of the table. It shows that what was deemed to be a bad balancer had a significant effect on the noise and was the greatest contributor at 2600 RPM (value 550). A balancer deemed good when assembled did not cause excessive noise, regardless of whether it was assembled to what was determined to be a good block or a bad block at 2600 RPM (value 10). The noise at the elevated 2,600 RPM greatly exceeded the noise at the lower 1400 RPM level (value 50) with the bad balancer.

In this case, we needed to control the balancer quality to meet or exceed the standard that was used in the testing that determined acceptable balancers.

In short, switching good and bad components or sections and measuring the noise output revealed the characteristic(s) in the good and bad blocks that turned out to be the most significant cause of the noise problem.

Cracked or Broken Problem Sheet

The sheet in Figure 8-11 can be used as a qualifier for studies involving cracked or broken items. It was developed over time after studying the conditions that led to numerous broken objects. The premise of the questions is to determine whether excessive energy was applied to an acceptable part or whether the strength of the part was insufficient to withstand normal forces. (You will find examples in Appendix D.)

Industrial Problem Solving Simplified | 151

Cracked or Broken Problem Sheet (Condensed)

Problem: _____
This problem type is caused by either inadequate part strength or excessive force applied. If it is a part of proper strength without any impact or blemish mark then suspect excess pressure, force or vibration effects.

Present: Desired:

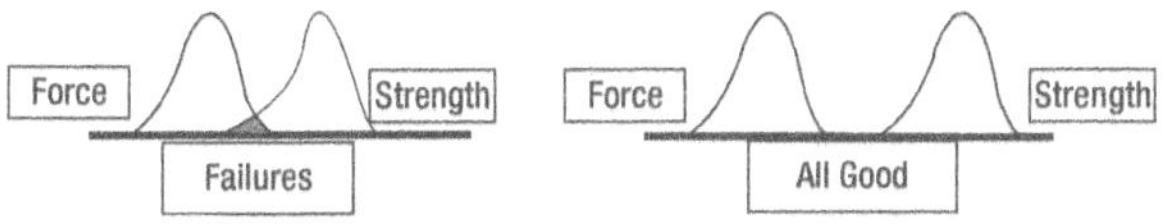

[Force] /\ /\ [Strength]   [Force] /\   /\ [Strength]
         Failures   All Good

What are the date codes? _____
What serial numbers are involved? _____
How many have occurred to date? _____
Is a new design or part being used? _____
Has a concept diagram been made? _____
Is there evidence of impact or other witness marks? _____
Do broken parts meet all specifications? _____
Has supplier made any changes? _____
Equipment, material or method changed last 3 months? _____
Is there a step or parting line at the crack area? _____
Is die mismatch present? _____
Any unusual patterns or marks present? _____
Is there only one flow path for the process? _____
Are you able to replicate the failure? _____
Do these parts have rework history? _____
Are handling drops greater than 2 feet? _____
Are parts clamped, pushed, or indexed? _____
Have push and clamp forces been verified? _____
Are there gaps in the conveyor system? _____
Are there any unusual transfers? _____
Is there any barrier interference? _____
Have parts fallen to the floor? _____
Do all the flaws occur in the same place? _____
Do all the parts fail after the same operation? ____
What else could cause this component to break? ___

Figure 8-11. Cracked or Broken Problem Worksheet

Not only is it necessary to elucidate the force-and-strength relationship, it is necessary to identify the individuality, severity, location, time, and relative frequency of occurrence. Figure 8-11 indicates some causes; add other items as you experience them.

When working these problems, while you're gathering individual identification factors, evaluate the energy-and-strength relationship. The top-left side shows that the part strength and the energy applied overlap in the area shown in gray. This indicates that there is the potential to have the components with the lowest strength overpowered by some of the highest forces applied. This may result in damaged or broken parts. We can surmise that if the energy sometimes exceeds the part strength, there will be damage some of the time.

Chapter 8 | Use Innovative Analysis Tools

The diagram on the right in Figure 8-11 shows that if the minimum strength of the component is never exceeded by the maximum force applied, there will never be damage due to excessive force.

Sometimes the component's weakness is so obvious that you don't need to conduct a complete analysis. Consider the following example, which involves a manufacturing problem caused by improper component strength. This condition occurred on a newly installed engine line during a pilot run when the components, methods, assemblies, functions, and final test were being certified to the final manufacturing assembly requirements.

While inspecting the first six engines that progressed down the assembly line, the operator was to check for obvious defects, such as oil leaks, poor connections, missing components, visible damage, and other detrimental conditions. He noted that when he removed the oil dipstick from the newly assembled engine, the dipstick was broken.

As part of its manufacture, the dipstick had holes punched into the end to indicate oil level. The crack always appeared at the center of the fourth hole from the end (see Figure 8-12). Because the breakage was noticeable after merely inserting and withdrawing the dipstick from the engine, it appeared that the problem was related to the dipstick strength and not to the forces that were applied to it. This was a valuable clue, because it is important to ascertain whether the part is weak or if excessive force has been applied.

Oil Dipstick: Always breaks across 4th Holes which are punched-out

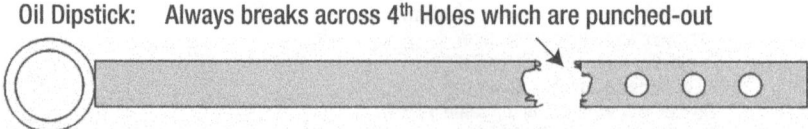

Figure 8-12. Broken Dipstick

The engine was not scrapped, but it did have to be reworked to remove the broken piece, which caused contamination and a throughput problem due to required rework. The recovered pieces from the crank case were pristine in that there were no metal slivers or abnormalities that could cause additional damage once the piece was removed. The engine was cleaned and reassembled.

There were no unauthorized witness marks on any of the broken or unbroken parts. None of the parts contained a die or serial number to differentiate handling or processing. All of the parts were broken in the center of the fourth hole from the end of the dipstick, which showed that there were differences between the bottom hole and the top hole, one end of the part to the other, and they were all breaking in one region in a recognizable pattern.

Industrial Problem Solving Simplified | 153

New parts were subjected to a bending test, and we found that the end opposite the indication holes could be bent 90 degrees more than 25 times without breaking, whereas the end with the holes cracked upon the first 90-degree bend. Comparing the end strength thus revealed that only the end with the punched holes cracked after bending.

We then compared the new dipstick to the older model, which had always been acceptable. The old dipstick had indicator dimples, not holes, to show the fluid level; it also was not crosshatched on both sides (see Figure 8-13). Obviously, the older design was significantly stronger than the new one. Therefore, the new design immediately became suspect.

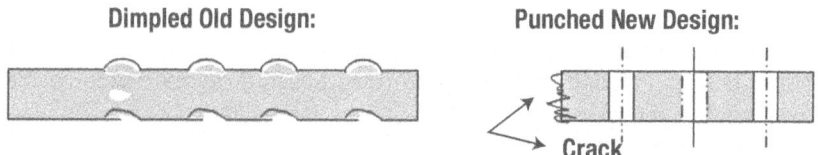

Figure 8-13. Dipstick Dimple versus Punch Design

Considering all the clues generated in this simple analysis allowed us to compile a summary of the significant findings. We found that the newly designed dipstick was deficient. The old design was dimpled and not punched. In addition, the crosshatching that was present on the new design appeared to be too deep and was believed to be a major contributor, if not the main reason, for the failure.

▪ **Tip** Replicate the defect whenever possible; it will usually provide valuable clues.

We had to stop the line and change the dipstick back to the dimpled design without the crosshatch. This resolved the immediate problem. However, that was not the end of the correction process. We revised the design (the DFMEA) to incorporate these findings so as to prevent a similar problem whenever another dipstick or similar indicating device is being designed. It also appeared obvious that the new design was not test-validated. We therefore also changed the PFMEA and control plan to ensure that this type of change would be subject to testing and validation next time around.

We changed the process (the PFMEA) to recognize knurling that had to be controlled on any future application. We also changed the control plan to validate any design change before acceptance and to test replacement components before use. The control plan also recognized and specified the favorable dipstick method and the need for visual characteristic requirements. Then we sent photographs to all concerned and captured the information for the reference file. This included the supplier who designed and furnished the component.

Summary

This chapter has shown you how visual inspections are not only helpful but almost mandatory when you're analyzing problems. Visual indicators—whether observing parts or the process, viewing the identification markings as shown by flowcharts and lot identification, or inspecting broken components for tell-tale markings or differences—can be useful in eliminating problems. Sound and noise studies can also aid your detective work. With these tools now in your tool packet, you'll learn about how to establish consistent work patterns and use many-level reviews.

CHAPTER 9

Establish Consistent Work, Many-Level Reviews, and Certification

Step Eight

Applying the principles of consistent work, many-level reviews, and certification constitutes the eighth step of the problem-solving process.

Consistent Work

Consistent work is a synonym for unvarying job performance. Clear and defined consistent work specifies the required operations and the method in which they must be performed. Clearly defined instructions are essential to ensuring

Chapter 9 | Establish Consistent Work, Many-Level Reviews, and Certification

quality performance. Posted job instructions allow employees to rotate jobs and to operate in cell arrangements while preventing faults.

■ **Tip** Post job instructions at every workstation to ensure consistent work. Posted instructions also help in manufacturing situations, where plant personnel rotate jobs or work in cells.

When you establish methods to prevent flaws and address the conditions that caused them, you become more effective. When problems do arise, make corrections as soon as they crop up. You should adjust any work instructions affected by operation mishaps because this enables you to apply corrections to current and future process operations. Consistent work also involves applying certifications to job instructions, procedures, and visual checks, as well as verifying that the changes you made to eliminate a flaw are in fact improving the process.

Many-Level Reviews

Many-level reviews enable you to appraise the operations process using multiple levels of assessment. These appraisals help to ensure compliance with consistent work procedures.

Reviews need not be overly complicated to be effective. They can simply be individual checks conducted by at least two or three pertinent people, such as the floor leader, the inspector, the supervisor, the department manager, or an outside assessor. Each of the assessors should observe the same condition at different random intervals. This allows for review of the process procedures as well as the effectiveness of the previous reviews.

These types of reviews are important, as they provide different levels of scrutiny. Any identified deficiencies can be corrected immediately. Many-level assessments ensure compliance with procedures because the reviews highlight any deviation from the expected work. Multiple checks provide redundancy and allow for certainty in the assessments.

An example of a preliminary many-level review is shown in Figure 9-1. In this case, some automatic bolt feeders became jammed in the customer automation handling equipment, which created line downtime. We created the checklist to prevent mixed bolts from being shipped to the customers. This checklist is rudimentary, and it can be revised as other problems occur.

Industrial Problem Solving Simplified | 157

Many-Level Review: Place a check mark after item found to be in need of correction.					
Determine if the following preventive actions are in place. Put an X by the items that are OK. Those not OK, make a comment.				Date:	
To Be Checked:	Leader	Inspector	Supervisor	Manager	Comments:
Damaged or leaking containers found and quarantined?					
Approved containers for parts storage?					
Only clean undamaged containers used?					
Empty containers checked before using? All old tags removed?					
Inspect containers for conditions that could allow part hang-up.					
All partially loaded containers covered when not loading?					
Skids loaded to safe levels to prevent spills?					
Containers in storage capped with cover?					
Identity tag installed on all containers before loading first part?					
Open storage containers eliminated?					
Components stored in covered shipping areas?					
Non-Conforming parts tagged and sequestered immediately?					
Area purged of parts before starting a new part sort?					
Dropped parts immediately picked-up?					
Output of system equals the input into the system?					
Work instructions updated after each incident?					
Parts removed for inspection tagged or identified?					
Parts left in inspection areas tagged and identified?					
Parts falling out of damaged containers or equipment?					
Parts lodged in containers, baskets, trays or conveyors?					

Figure 9-1. Many-Level Review Checklist Example

Chapter 9 | Establish Consistent Work, Many-Level Reviews, and Certification

In this case, the items were checked on a daily basis by at least a floor leader and an inspector. A supervisor or manager checked for these conditions on a weekly or monthly basis. An upper manager assessed compliance on a yearly basis. If anyone found any discrepancies, they were addressed immediately. New, detrimental items were added to a revised many-level review list for future assessments.

This checklist offers a preliminary many-level review that will help prevent mixed parts from being sent to the customers (internal or external). Although a mixed-part problem may seem inconsequential, it can cause jams in customer automation that can result in expensive line downtime. You should customize this checklist for your situation. You can also update it whenever you find another suspect condition. Since each manufacturing location has its own characteristics, that may require additional observations. In addition, any recognized problem can be added to the list and used in other applicable areas of the facility.

You can perform the entire review in a few minutes by simply walking through the area under appraisal and making the observations as listed. Once any other deficient circumstance is noted it should be corrected and added to future reviews. All items found deficient should be subject to immediate corrective actions.

These types of checks can be especially useful in maintaining established controls. They can be used to identify items and aid management in establishing new requirements for future manufacturing and assembly (PFMEA) and control plan requirements. These checks can also be used to ensure compliance with established procedures, services, and items such as:

- Protective equipment use
- Gauge calibrations
- ISO and quality systems requirements
- Housekeeping
- Label issues
- Routing procedures
- Rework procedures

Labeling

Proper labeling requires consistent work and many-level checks to ensure that labels are applied correctly and to the customer's or regulatory requirements. Even though labeling does not add a function to the part or assembly operation, it is very important for descriptive value. Any process that affects the labeling operation should be included in the job instructions that specify the consistent work as well as the checks that are being performed on the labeling operation.

How do you come up with labeling-related items for your many-level reviews? Let's look at an example.

As we've seen, correct labeling is very important in the manufacturing setting. This list of eight items can help you create assessment items for the purposes of label control. These items may be generated from problems experienced in the past or from items from a concept sheet.

1. Labels should be discussed at a quality planning or design meeting before the product is launched.

2. Label problems and processes should be addressed in the DFMEA, PFMEA, job instructions, and control plan documents.

3. Procedures must be set up for label storage, application, and control.

4. Employees should be trained in the correct labeling methods.

5. Review items should be ascribed from the conditions above. Each item recognized in any of the above activities should be included on the many-level checklist if it can have a detrimental effect on the finished product.

6. Problems must be corrected to prevent label problems when they are recognized.

7. Someone must be assigned specific responsibility for each of these steps. Each step does not have to be assigned to one individual. But assignments and the many-level checks should be made by specific individuals to ensure compliance.

8. The items identified as important should be added to a many-level review sheet for compliance. They should be added, as appropriate, to other lists used in other areas of the facility.

This "list of eight" is in fact a solid blueprint for carrying out any kind of assessment process. You can follow these steps to come up with an effective and timely many-level review checklist regardless of the process—it works with machining processes, first-piece inspections, assembly instructions, damage control issues, reworks, mixed parts, foreign material issues, shipping processes, and communication problems.

Chapter 9 | Establish Consistent Work, Many-Level Reviews, and Certification

Advocates: Assign a Manager

Once you have procedures and controls in place for each process, your organization can quickly attack any defects that arise. Where are problems most likely to come from? As an example, let's look at the data from a high-volume manufacturing facility. During a specific time period, there were 513 flaws noted. We classified these faults as follows:

1. Assembly   186
2. Machining   127
3. Communication/shipping   94
4. Damaged components   45
5. Mixed parts   22
6. Foreign material   20
7. Label issues   19

Total = 513

Each of these faults is an opportunity for improvement. To be effective, there should be an advocate—generally a manager—assigned to each major area of responsibility. In addition, there should also be an alternate assigned for when the advocate is absent or unavailable.

Assigning an advocate is not compulsory. However, if everyone is responsible for controlling a problem, you will likely find that no one has accepted that responsibility. Conversely, when advocates with responsibility find discrepancies, they have the authority to apply fault-removing corrections like improving the design or process FMEA, adjusting the control plan and consistent work procedures, and modifying the job instructions and training process. Together, these corrections provide a robust quality-assurance system.

In addition, having an advocate will expedite problem correction because someone is held accountable. Once it's assigned, the task can be measured and fulfilled.

Certify Problem Solutions and Verify

Now suppose that you have a manufacturing problem, and you have reviewed all the available information and collected the pertinent data. You are about to apply an action that you believe will correct the problem. However, suppose there is a nagging sensation in your mind that's causing you some trepidation about the recommended solution. In that case, you should talk it over with your supervisor to enlist their aid or buy-in. It's possible that unless you act, no one will solve the problem.

To prove or verify your trial solution, consider using Figure 9-2, a flaw chart. You can use such a chart to outline any manufacturing operation that has variability. It can be an effective tool in communicating the past, current, and potential effect of a process. Charts of this nature are stalwart certification tools.

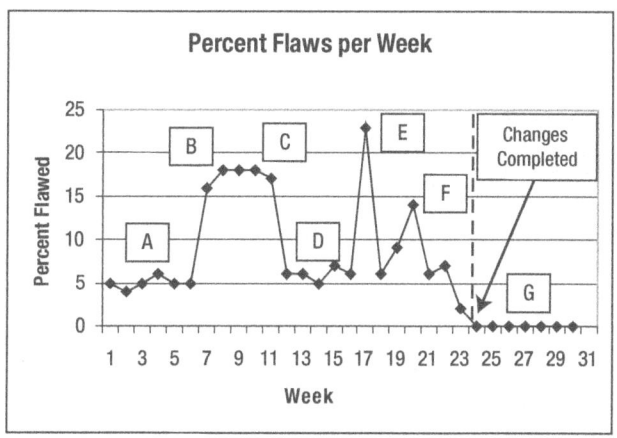

Figure 9-2. Chart Showing the Percentage of Flaws per Week

Initially, the process was limping along with a flaw rate of 5 percent, as indicated by the A in the chart. Around week 8, the defect rate shot up to 18 percent, shown by point B. At this point, management becomes aware of the problem and attempts to fix it, as indicated at point C. While the process is being investigated and nothing substantial has changed, the process returns to a 6 percent flaw level, as indicated by point D. At this point, unfortunately, management thinks the problem has gone away. In most cases, manufacturing will continue to run at this level, with intermittent spikes caused by chance, variability, or unknown interactions, as indicated by points E and F. Until the problem is identified, it will not be solved.

You look more closely and notice that there are erratic flaw surges during the 7th, 17th, and 20th week of the production run.

By this time, you have already performed an analysis that leads you to believe that you know what is causing the problem. Maybe you used a comparison of five duos or compared five good with five bad samples to both discover and verify the problem.

Chapter 9 | Establish Consistent Work, Many-Level Reviews, and Certification

You make a change based on your study and then compare five of the previous samples with five of the new ones. You find that your change has significantly reduced the flaw rate, as shown by point G. This is why it is important to record the changes and their results—you can track and verify their results and then potentially incorporate them elsewhere in the facility.

> **Note** To ensure that a change is satisfactory, some proponents suggest that the old method be tried one more time so that the flaw is recreated as a temporarily trial. I do not advocate this approach, as it makes management very nervous and can be wasteful. Rather, use a chart like the one shown Figure 9-2 to track the effects of any changes.

During the 24th week, the corrective changes were completed. The percentage of flawed product was reduced to zero for the next seven-week period. The adjustments you made thus corrected the erratic surges as well as the recurring low-level flaws.

The graph in Figure 9-2 complements the information provided by the comparison of duos and the comparison of five good versus five bad, which in each case provided over 95% confidence that there was a significant difference in the effects of the characteristics being compared. Afterward, a chart like 9-2 compared the changes to the flaw level in the five bad weeks prior to the change and the five good weeks after the correction. The chart confirms with 95% confidence that the changes improved the process. There is clearly better performance in the five weeks after the change than the five weeks before the change. And there is no data tie or overlap between the two periods.

Case Study: The Value of Comparisons

The following example—in summary form and without the study data—shows how helpful it is to use the duos and the comparison of five good versus five bad to verify whether a process change was successful.

The supplier in this case study creates a certain chemical for its customers. Its process involves using bulk chemicals, mixing vats, processing towers, and a packaging and shipping operation. During the materials processing, the chemicals are mixed and transported through the piping system to the individual towers, where additions and reductions in the process are made. The towers consist of three stages and include temperature control to activate the chemical process.

Now the problem. The process jammed up at random times and had to be shut down for cleaning. This downtime resulted in production loss and in excessive labor costs to clean and restart the system. The downtime also generated losses due to waste of the materials that had been incorrectly created.

Over time, they found that adding bags of lime could sometimes prevent the jams. So they limped along creating what they could, adding the lime whenever the system started to act up. However, they continued to experience erratic downtime and excessive costs.

The supplier collected data on 17 variables over a period of one month. These variables ranged from the day's production in tons, feeds and speeds of material inputs, air temperature, humidity, bags of lime added per hour, moisture content of the lime, tower temperature in each stage, and final analysis of the product.

The data did not reveal a clear plan for correcting the problem. There were no apparent clues that indicated what was creating the process jams. The problem-solver assigned to solve the output and jam-up problem didn't use any of the methods described in the previous chapters. Because of this, it appeared that the main data point to be used were the five days with the most production output (the best five). There was a significant difference in the tons of output generated each day. This information was entered into a spreadsheet in descending order. The other variable information that was collected was entered in relevant columns. The table was pretty, but it didn't reveal any significant findings. Some numbers were very similar and some varied considerably. The spreadsheet offered no real action points.

Then, with a stroke of luck or insight, they decided to list all the variables for the worst five days of production when a jam did not interfere with the operation. (The worst five days that included a significant jam up were not incorporated into the spreadsheet, because that information deteriorated or was not available. Once the process had a jam-up problem, the data was unable to be generated because the process required that it be stopped.)

The chart of the five bad days showed the same variability as the five best days. With the exception of the ranked daily output in tons, the other variables appeared at first to be random.

Only after combining the two sheets and listing the total production output for all ten days in descending order did the a pattern emerge.

The only major relationship present in the data suggested that when the temperature in Tower 2, Zone 2 was kept at a higher temperature, the production tonnage increased.

At first, they did not fully appreciate this clue. The information was brought to the attention of the manufacturing supervisor, who decided to conduct an experiment. There had been no temperature specification for Zone 2 in Tower 2. So randomized production runs were conducted with the temperature in Zone 2 classified as a critical control. Two distinct temperatures were used.

Chapter 9 | Establish Consistent Work, Many-Level Reviews, and Certification

The supplier chose a five-day test period for each of the two types of temperatures. After test randomization, high and low temperature data was collected for the days. The days with the higher temperatures had significantly higher production tonnage output than did the days with the lower temperatures. The sum of extremes count was 5 + 5 = 10, which was significant. In addition, the use of lime was greatly reduced when the higher temperatures were maintained.

Many people were overjoyed, but there were a few in management who worried that this was just a statistical anomaly. To add creditability to the results, the supplier plotted the production output on a daily basis. After using the higher temperatures, production output rose significantly. The chart spoke for itself. Production had increased with the change in temperature.

This case shows that there is another useful way to verify whether a process that you have tried to improved has actually improved. Compare the bad five days before the change was made to the five days of production after the change was made. If the sum of the extremes is equal to 7 or more, you will have 95% confidence that you have made the correct change. This is further explained in Appendix E. It explains how to use the sum of extremes test for testing and verification.

Tip There is only one other little thing that you might want to consider. When you attack a problem, take the time to discuss what is happening with the people in manufacturing who are actively engaged with the daily frustration. I don't mean the superintendent or the supervisor, I mean the individuals who have to perform the manufacturing operation. These people have insight and special knacks for performing the operation to make their own jobs easier. They have a lot to offer when it comes to solving problems.

Summary

Now that we are almost at the end, I confess that I have tried to improve on an old teaching axiom to make the information easy to understand and use. After you become proficient in these practices, please consider using the same approach with your peers that I have attempted to use with you. It is as follows:

- I told you that there were simple useful tools that could help.
- I provided examples and showed you how they operated.

- I explained the examples so you understood what you were seeing.
- I instructed you to use the tools that are available to you.
- I repeated as many times as I could what you had been told.

I strongly recommend that as you get better at solving problems using the tools presented in this book, you also use this approach to educate and inform your peers. Industrial problem solving can be simplified.

I believe that you now have all the information you need to quickly and easily generate clues by identifying important process characteristics. By using the clues generated by comparing the data that you accumulate, you will be able to identify the suspect variables that are contributing to the problem. In addition, using concept drawings and check sheets will make you better at identifying any conditions that are present. Once you identify the adverse conditions, you can propose solutions. After you enact the corrections and verify that the process is satisfactory, you must verify the corrective action. With each project investigation, you will find these methods easier and easier to apply.

So let's move on the book summary.

CHAPTER 10

Summary

In summary, the following eight steps are required to solve problems simply and successfully:

1. Adequately and completely define the problem.
2. Define the fault characteristics.
3. Construct a concept sheet to augment a plan of attack.
4. Develop a plan of attack.
5. Collect accurate data that is relevant to the study.
6. Choose and use analysis tools (observations).
7. Use innovative analysis tools (easy-to-use comparisons).
8. Establish consistent work and many-level reviews.

Preliminary requirements for eradicating faults are as follows:

- Agree on the specific problem definitions.
- Be sure that everyone involved is able to visually inspect the fault, and the operation that produced it. If this is not feasible, provide everyone involved with a photograph of the fault that allows it to be studied for size, shape, and other criteria.
- Verify that none of your suppliers made changes to the product, either to the materials or to the process, no matter how slight.
- Specify the problem considerations on a concept diagram.
- Take the time to look at the process from beginning to end.

Chapter 10 | Summary

- Determine if the problem occurs in a specific area. Trace it back to the last acceptable operation or flow path operation.
- Agree to the nature of the fault. Is it due to poor design, poor specifications, part weakness, excessive forces, abuse, mismatches, or malfunction?
- Determine if there are any witness marks or clues.
- Determine if there is one best or worst clue by pattern, serial, mold, machine, operator, shift, day, line, location, mixture, method, and so on.
- If the fault can be reproduced, try to replicate it.
- Decide if the fault is due to inadequate strength or excess energy.
- Capture at least five good and five bad samples for comparison.
- Generate the visual clues and analyze the data that you find.
- Ensure that only approved material and process changes have been made.
- Assign advocates to manage and guide the new system.
- If your supplier has more than one instance of the same malfunction, have them certify the product that they send you.
- When all else is ineffective, revisit the scene of the crime.

If you follow these methods, your mind's eye will quickly learn to absorb and analyze clues to the solution. Good luck!

APPENDIX A

Fractional Explained

You can use fractional analysis (a partial factorial) to determine which variables are important. Fractional analysis can designate variables to be used in designed experiments. It can also be used to generate clues when other methods have not worked.

This appendix explains fractional use in more detail than Chapter 7. It explains and describes the required calculations. These calculations are not difficult, and any investigator will become proficient after creating one or two fractional.

If you don't recall the problem presented in Chapter 7, review the piston-stuffing data sheet and the resulting piston-stuffing interaction table, shown in Figures 7-19 and 7-20. These illustrations show how we accumulated the data. As you'll see, I highlighted the calculations that focus attention on the parameters being calculated.

Figure A-1 illustrates the calculation of the effect of two variables on a piston-stuffing operation, as described in the main text. The operation consists of two stuffing machines, which insert two pistons each into a machined engine block.

Appendix A | Fractional Explained

The table is a fractional in a 2X2 layout. It was designed from the data collected when the machines were jamming. We identified at least three main variables from the data. These included the two stuffing machines, the quality of the pistons, and the two different machining lines for the casting blocks.

The first display is shown in Figure A-1. Variable 1 compares a stuffing operation done in Operations (Op) 2050 and 2060. Variable 2 is the machining experienced when processed through Stream A or Stream B. The dependent variable represents the total number of engine blocks that resulted in stuffing failures. These failures were recorded for each of the four classification combinations during the data-collection period. Calculations are as follows.

For the operation effect: The sum of failures for Op 2050 for the A and B machining operations was 4 + 1 = 5. We divided this number by 2 and inserted it in the *Xbar* (average) block to the upper right of the matrix. This indicates that the average effect of Op 2050, when stuffing A and B machined blocks, was 2.5. The effect of Op 2060 was determined by adding the failures: 23 + 2 = 25. We divided 25 by 2 and inserted the result—12.5—in the highlighted lower-right Xbar block. It shows the effect of Op 2060 when it's running A or B machined blocks. The difference between these two averages is the operation effect of the difference of the two stuffing operations: 12.5 − 2.5 = 10. This is the farthest right value shown in the figure.

Piston Stuffing Fractional

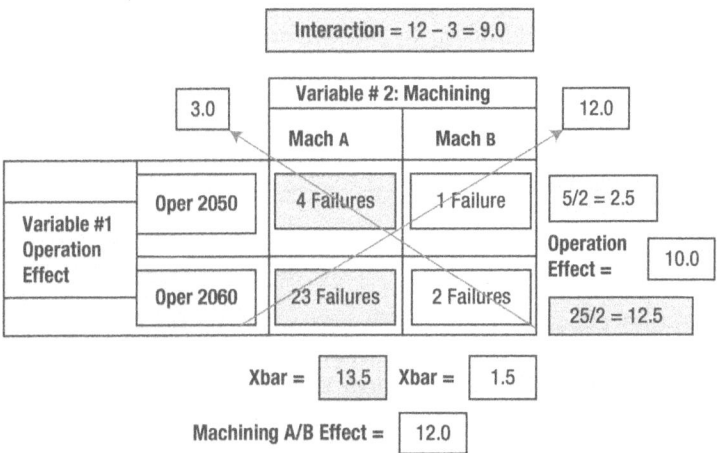

Conclusions Based on Current Sample:

Machining A caused the most failures with an effect of 13.5.

Op 2060 caused the second-highest variation with an effect of 12.5.

Machining A and Op 2060 have an interaction with an effect of 9.0.

The direct, indirect, and interactions are strong with Op 2060.

Something is wrong at the Op 2060 stuffing station.

Machining effects are larger than operation-stuffing effects.

Recommendations:

Machining A must be refurbished and retargeted to nominal dimensions.

Stuffer at Op 2060 must be checked for compliance.

Figure A-1. Fractional of Operations and Machining with Conclusions and Recommendations

For the machining (A or B) effect: The sum of failures for blocks machined by A for the Operations 2050 and 2060 was 4 + 23 = 27. We divided this number by 2 and the result—13.5—and was inserted and highlighted in the Xbar block in the bottom left of the matrix. This indicates that the average effect of machined A blocks stuffed by Op 2050 and 2060 was 13.5. The effect of machining B was found by adding the failures (1 + 2 = 3). We divided this number by two and the result (1.5) was inserted in the lower-right Xbar box to indicate the average machining B effect. The difference between the two Xbar effects was 13.5 − 1.5 = 12.0 for the total A versus B machined effect on the system. This value is shown as the lowest boxed value in the figure as the machining A /B effect. This is the largest total effect and is therefore the most important difference to be studied for correction.

We made similar calculations for the diagonals to determine the interaction effects of the variables, described in a following paragraph.

The interaction effect was 9.0, which was smaller than either of the direct effects of machining A or stuffer 2060, but almost as detrimental. This value is highlighted at the very top of the figure. It represents a comparison of the two diagonal value sums, shown as arrows. The diagonals were calculated as follows: Op 2050 had four failures with machining A and Op 2060 had two failures with machining B. Therefore, (4 + 2) / 2 equals 3.0 for the diagonal box to the top left. Similarly, Op 2050 had one failure for machining B and Op 2060 had 23 failures for machining A. This resulted in a diagonal score for the upper right of (23 + 1) / 2 = 12.0. The interaction was then the difference, calculated as 12.0 − 3.0 = 9.0.

From this, we can see that the largest matrix effect was due to the difference between machining A and machining B (the machining A/B effect, 12.0). The second largest matrix effect, the difference in the operation effect between Op 2050 and Op 2060, was 10.0. The third largest matrix effect was due to the interaction between the operations and the machining differences, and it was 9.0. However, machining A had the largest variable effect with a 13.5 rating, which indicated that it was the most important variable and was the candidate for being the main cause of the problem.

These results indicate that there are at least two and possibly three conditions that require correction. Actually, looking at the 2 X 2 matrix, you might suspect that there are obvious clues. So what does the data tell us when we combine and analyze all of the paired combinations and classifications in Figure A-1? The following facts become obvious:

- Most of the defects in both operations are from machining A. Machining A had 27 total jams, versus three for machining B.

Industrial Problem Solving Simplified

- Machining B blocks run through both operations with a minimum of jams. Machining B had only three jams of the 30 total.

- There is a difference in the stuffer operation or setup because machining A blocks do not run through Op 2050 and Op 2060 with the same problem frequency. Op 2050 had four jams versus 23 jams in Op 2060.

Again, as a reminder, I encountered this problem on an assembly line when the process deteriorated and started jamming, which caused downtime and lost production. We used a data sheet to collect data, as described in the main text, and we came to the conclusions based on the data, which pointed to the three variables used in the calculations.

We identified actions, indicated on the display sheets, that led to correcting the problem causes:

- Machining A must be refurbished and retargeted to nominal dimensions.
- Stuffer at Op 2060 must be checked for compliance.
- Piston quality is not suspect for the current jams.
- Retarget machining A for bore 2 to nominal dimensions.
- Retarget machining A for bore 4 to nominal dimensions.

Similar calculations are provided in Figures A-2 and A-3 for all the other variable combinations. They show the direct and interaction effects of the bores and pistons included in the system. These three variables had the most significant effect on the equipment jamming on the assembly line. The original suspect, defective pistons, is the cause of least concern.

Appendix A | Fractional Explained

Piston Stuffing Fractional

<u>Stuffing Reject Failures Using Good and Suspect Pistons with Machining A and B</u>

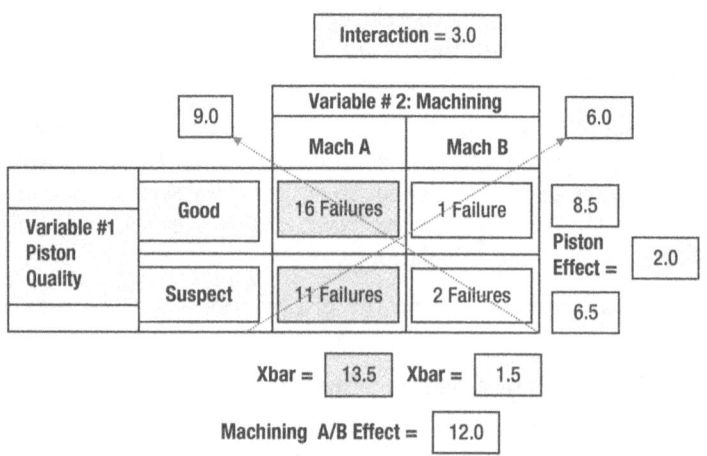

Conclusions Based on Current Sample:

Machining A caused the most failures with an effect of 13.5. (Repeat).

Good pistons had more rejects than suspect pistons at an effect of 2.0.

There is interaction between machining A and pistons at an effect of 3.0.

Interaction has greater effect than direct effect of pistons alone (3–2=1).

Machining A causes the most defects overall.

Recommendations:

Machining A must be refurbished and retargeted to nominal.

Piston quality is not suspect for the current problem jam-ups.

Figure A-2. Fractional of Pistons and Machining, with Conclusions and Recommendations

Industrial Problem Solving Simplified

Piston Stuffing Fractional
Operation # 2060

<u>Stuffing Rejects Using Comparison of Machining A, Bores 2 and 4, and Operation # 2060 with Use of Both Good and Suspect Pistons</u>

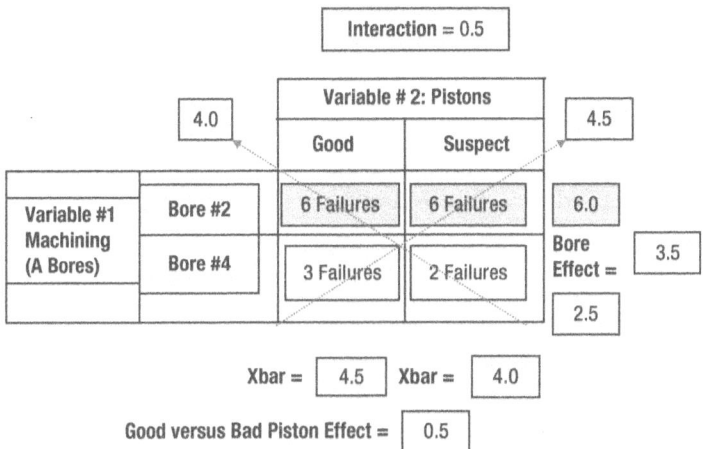

Conclusions Based on Current Sample:

Bore 2 is responsible for most reject variation with effect of 6.0.

Bore effect had 7 times more variation than pistons which had effect of 0.5.

Bore effect had 7 times more variation than part interaction effect of 0.5.

There appears to be no difference in piston characteristics for this problem.

Recommendations:

Retarget machining A for Bore 2 to nominal dimensions.

Retarget machining A for Bore 4 to nominal dimensions.

Figure A-3. Fractional of Pistons and Block Bores with Conclusions and Recommendations

Appendix A | Fractional Explained

As a review, the three figures required the following actions:

- Machining A must be corrected to the target dimensions (see Figure A-1).
- Op 2060 stuffer must be checked for target dimension (see Figure A-1).
- Pistons are not significant in causing jams (see Figure A-2).
- Machining A must be retargeted for bores 2 and 4 (see Figure A-3).

We solved the problem by performing these tasks. The purpose of this example is to show you why fractionals are important and easy to use.

APPENDIX B

Interaction Explained

An *interaction* can best be defined as a violent reaction that interrupts normal conditions. Interactions become apparent when there is an intense outcome created by chemical, thermal, or physical responses to the combination of two or more variables.

If a flaming wick is inserted into a combustible gas stream, for example, there will be a flash as the gas ignites. This is an interaction. See Figure B-1.

Stuffing Example

Two or more conditions must be present to cause an interaction, as in the case of the piston-stuffing example discussed in Appendix A. Machining A was to specification but not at the nominal dimension. The stuffer station, 2060, was operating satisfactorily with blocks processed by machining B but was jamming with blocks processed thru machining A. This is a detrimental interaction of the machining A and stuffer 2060 operation.

You might understand the idea of an interaction better by viewing the next two examples. In the first example, the holes in the flange will not strip if the threads are good, even when the flange tensile strength is low. Conversely, when the threads are poorly formed, they tend to strip is significantly greater as the threaded tensile strength declines even though the flange material is of acceptable chemistry and strength.

Appendix B | Interaction Explained

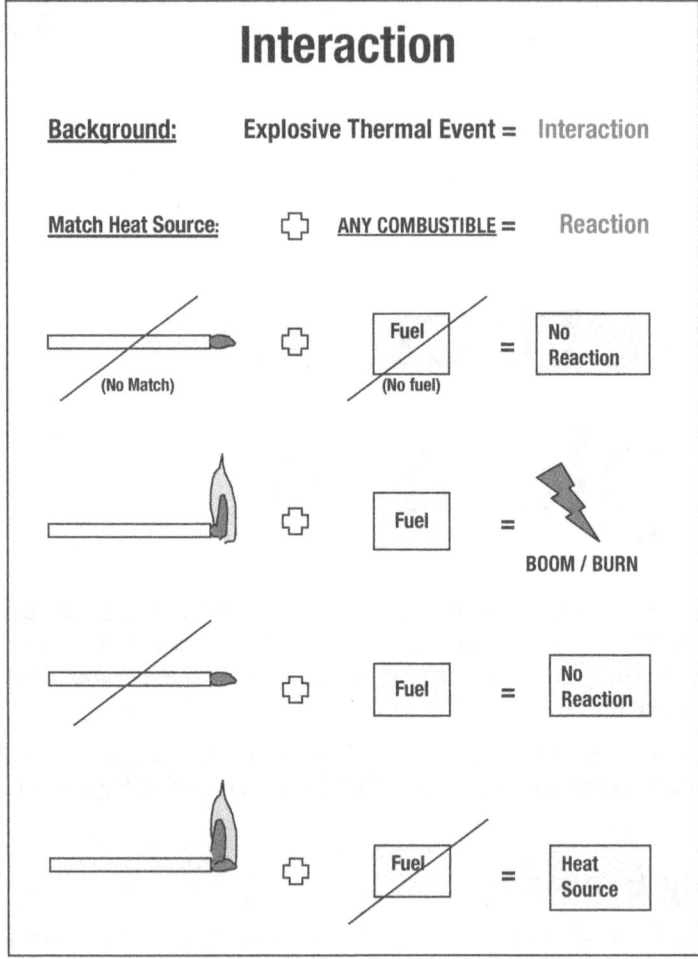

Figure B-1. Interaction Concept Sheet

Stripped-Threads Example

There are variables that can affect threaded flanged component tensile strengths. If there is variation in the flange-manufacturing process, it may be possible to get unacceptable flange tensile strengths at some time. A threaded flange may not strip if the threads are completely formed in either an acceptable or unacceptable tensile strength flange. However, if there is a weak tensile strength flange *and* the threads are not completely formed, the threads will strip and fail. This is also an interaction (see Figure B-2).

Supposition:

If there had been a relatively few failures, and then a huge spike in flaws, there may be an interaction present. If the failures were just due to material strength, only some of the defective batch would generate stripped bolt-hole failures. Since there were only a few failures and then a large spike, there may be another variable present causing an interaction.

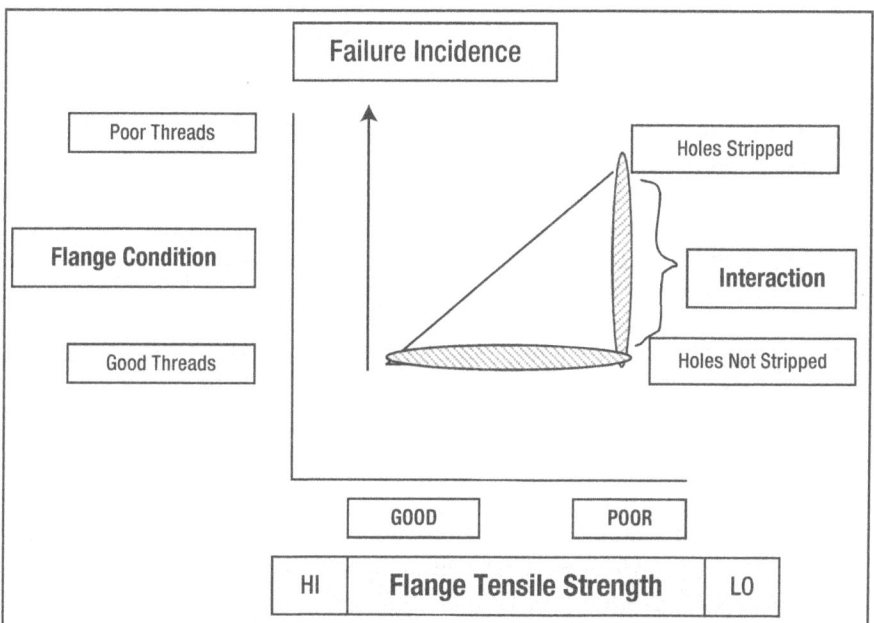

Conclusion:

Threaded bolt holes will not become stripped in the flange if the threads are properly formed. If the threads are not formed properly, they will not provide sufficient tensile strength to support the force of the bolted assembly without failure. The incidence of only a few failures in a large lot reinforces the conclusion that the threads are properly formed. Therefore, a huge spike in flange bolt-hole failures may indicate another variable in addition to chemical property strength may be present and interacting. See the Appendix A for method to calculate interactions.

Figure B-2. Interaction Diagram and Analysis

Balance Shaft Assembly Noise

The next example is a study of balance shaft assembly noise. It shows the data that was collected at two different RPMs with both noisy and quiet engine cases and balancer shafts. Figure B-3b indicates where an interaction between the balancer and the block at two different RPM levels happens. You obtain the ranking data for an interaction by subtracting the averages of the diagonal additions. This value is important when it exceeds the direct effects caused by the individual variables in the matrix. (Refer to Appendix A for means to calculate interactions.)

To begin understanding interactions, you must first realize that two individual components or processes may be individually acceptable. However, when two—we'll call them *sympathetic*—conditions combine to form an unusual and potentially violent response, we have what is known as an *interaction*. Simple examples of interactions are as follows: A combustible gas captured in a container is safe and stable unless a match or flame source is applied, which then results in an explosion or at least combustion. This is an interaction. Another example was the violent explosion mentioned earlier, caused when molten iron was spilled on a damp floor. The encapsulation and sudden release of the steam trapped by the molten iron was an interaction. Similarly, the mating of two previously compatible components may result in an interaction during mixing or use.

In this case, we noticed that some engines suddenly appeared to be noisier than other produced engines while under testing. We developed a study to determine and eliminate the causes.

Figure B-3a shows the results of the testing, which involved what were considered good and bad engine blocks and good and bad balancer components. "Good" indicated quieter blocks and "bad" indicated noisy ones. Using the same type of calculation of the diagonals explained in Appendix A, as shown by the arrows, we calculated the conditions that caused the most noise (see Figure B-3b). The assembly combinations were tested the 1,400 and 2,600 RPM. At 1,400 RPM, the interaction effect was shown with a difference of 35. This was the difference between the two diagonal measurements: 120/2 = 60 and 50/2 = 25, and 60–25 is 35. The data at 2,600 RPM indicated an interaction difference of 145, but there was also a more significant direct effect difference of 405 when a bad block was used. This is observable on the upper-right display.

You can see that interactions can be present not only in explosive forces but in underlying process variations. You can also see, from the two lower charts in Figure B-3b, that the RPM also had an effect and that it was dependent upon either or both a bad balancer and a good balancer. The difference of 240 for the bad balancer and bad block was almost as detrimental as the independent bad block effect of 405. The block machining was retargeted and only acceptable balancers were used in the ensuing production.

Hopefully, these examples provided a clear understanding of what an interaction is and how it can affect a process. This can be important because, unless you look for them, they will be difficult to identify.

BALANCE SHAFT ASSEMBLY NOISE STUDY

RPM	GOOD Block & BAD BSA	GOOD Block & GOOD BSA	BAD Block & GOOD BSA	GOOD Block & BAD BSA	BAD BSA to BAD BSA	BAD BSA to BAD BSA	GOOD BSA to GOOD BSA	GOOD BSA to GOOD BSA
	RPM vs RPM	RPM vs RPM	RPM vs RPM	RPM vs RPM	BAD Block	GOOD Block	BAD Block	GOOD Block
1400	70	50	30	20	70	20	30	50
2600	600	40	300	50	600	50	300	40
Difference	530	10	270	30	530	70	270	10
	3rd				2nd			

RPM	BAD Block To GOOD Block - BAD BSA	BAD Block TO GOOD Block - GOOD BSA	BAD BSA To GOOD BSA With BAD Block	GOOD BSA With GOOD Block
1400	50	30	40	30
2600	550	20	300	10
Sector Difference				
	1st		4th	
	5th		6th	

Findings:

1st Block to block differences are the most important to be evaluated (largest variation = 550)
2nd RPM differences contribute to 2nd largest family of variation (variation = 530)
3rd BAD block and BAD BSA combination assemblies are influenced by RPM (Variation = 530)
4th Differences between GOOD / BAD BSA do not significantly affect the variation at different RPM in GOOD blocks
5th RPM is part of the interaction as largest family of variation (1st) is affected significantly at the different RPMs.
6th RPM interaction is not present or minimal with GOOD blocks.

Conclusion: Must find clear data separation in characteristic that interfaces with BSA in GOOD / BAD Blocks.

Figure B-3a. Balance Shaft Assembly Noise Example

Appendix B | Interaction Explained

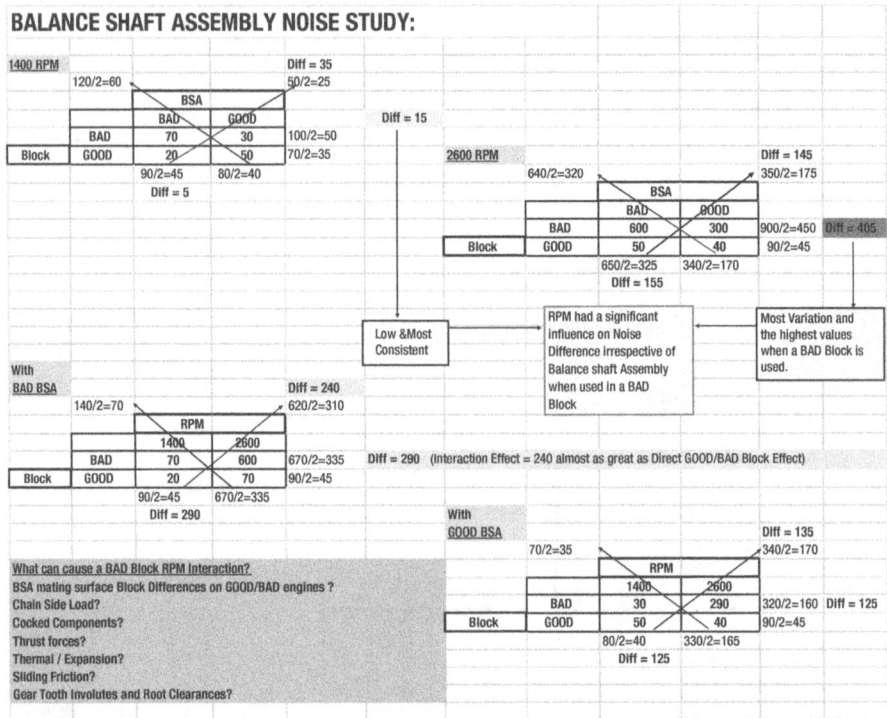

Figure B-3b. Balance Shaft Assembly Noise Example with Interactions Shown

APPENDIX C

Cracked-or-Broken Example

This appendix contains more examples of cracked-or-broken problems that we have been encountered in various industries. These examples are intended to provide you with other perspectives that you can use when you're evaluating cracked or broken components. It also explains why the cracked-or-broken check sheet included here should be used to help generate clues.

You can use the sheet shown in Figure C-1 when you're evaluating a damaged or broken component in conjunction with a concept diagram. The sheet is self-explanatory; questions relate to the strength of the component and the conditions that may have acted upon the component to cause failure. As you gain experience with these forms and methods, you will recognize other valuable considerations applicable to your products. You should save these recognized considerations or items of interest and incorporate them into your most useful sheets.

Cracked or Broken Problem

Problem: _____

This **Flaw** is caused by either: **Inadequate Part Strength** OR **Excessive Force Applied**
If **Proper Strength part with no IMPACT Mark**, then suspect **Excess Pressure, Force or Vibration**.

Present: Force / Strength — Failures
Desired: Force / Strength — All Good

What are the date codes?
What serial numbers are involved?
Has a concentration diagram been made?
Have flaw pictures been distributed?
How many have occurred to date?
These conditions to be investigated and explained on reverse side of sheet:
Do cracked surfaces show oxidation, paint or other inclusion?
Is there evidence of impact or other witness marks?
How much force is required to break it the same way?
Have you replicated the failure yet?
Do broken parts meet ALL specifications?
Are there: Rough handling by personnel?

EQUIPMENT:
- Drops of greater than 2 feet?
- Holes causing jam-ups?
- Gaps that can catch parts?
- Holes/cracks in transfer equipment?
- Gaps in conveyors?
- Have parts fallen to the floor?
- Barrier interference?
- Unusual transfers?
- Was rework involved?
- Are parts pushed or indexed?
- Are parts clamped?
- Have push/clamp forces been verified?
- Is push/crack/transfer area flat and smooth?

Is there only one flow path for the process?
Has equipment changed within last 3 months?
Has material changed within last 3 months?
Has method changed within last 3 months?
Has design been changed last 3 months?
Has personnel changed within last 3 months?
Has supplier changed within last 3 months?
Has location changed within last 3 months?
Has equipment been repaired within last 3 months?
How many times has equipment jammed last 3 months?
Is there a step or parting line at the crack area?
Are stress risers present at the crack site?
Any unusual patterns or marks present?

Figure C-1. Cracked or Broken Criteria

Industrial Problem Solving Simplified | 185

The two concept diagrams shown in Figures C-2 and C-3 can provide you with a better understanding of conditions that lead to cracked or broken components. You can use these conditions to analyze similar problems.

Concept Diagram

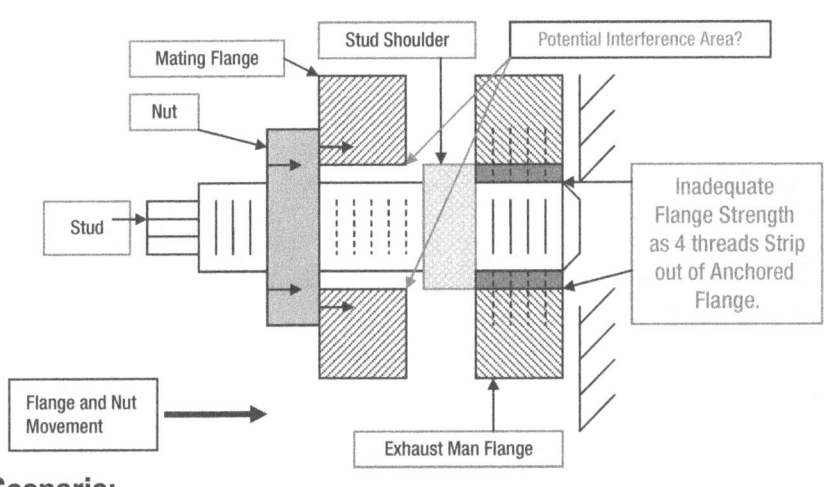

Scenario:

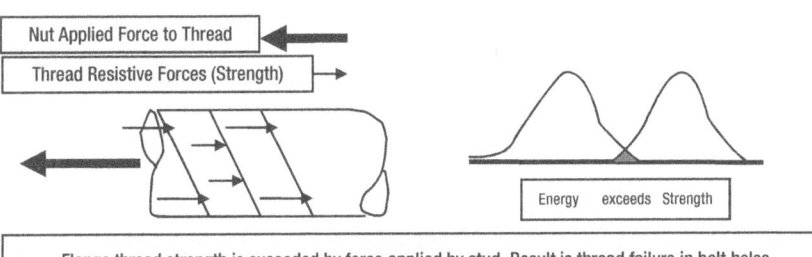

Flange thread strength is exceeded by force applied by stud. Result is thread failure in bolt holes.

Observations:

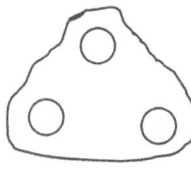

- No Failures in the top flange bolt hole to date.
- 3 failures in bottom left hole
- 2 Failures in bottom right hole
- Thread formation appears very rough with deformations and rolled threads visible.
- Took > 90 Newton Meters to strip sample of 2
- 2 different nut suppliers are involved
- Flange threads weaker than stud threads.
- Bolt holes accepted thread gage satisfactorily.

Figure C-2. Stripped Flange Bolt Holes

Appendix C | Cracked-or-Broken Example

Cracked Exhaust Manifold Concept Diagram

Problem: Exhaust manifold cracks at flange bolt hole.

Condition: Force — Cracked Manifolds

Suspect: Hot crack
Impact
Stress
Alignment
Excess energy applied

Part Configuration

Crack Propagation:

Dings found in crack area

Potential Force Generators
Impact against hard surface creates fracture/stress-riser   (Primary Suspect)
Off center to mating part
Hole misalignment and configuration
Sideway pressure of fastener
Tightening sequence
Hole/fastener Size
Flange/hole damage

Mating hole non-alignment

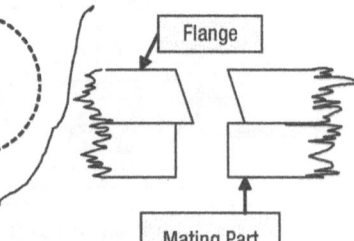

Flange

Mating Part

Findings:
1. Crack is an event as only a small % are cracked.
2. The energy applied exceeds the strength of the parts that cracked
3. Part strength is not suspect as chemistries appear to be in bounds.
4. **Impact ding after machining found on flange which created fracture plane.**

To Do:
1. Check for oxidation in crack to eliminate hot crack possibility.
2. Observe failed and mating parts for witness marks.   (Found Dings on 2 of 4)
3. Get information on tightening sequence and forces.
4. Get capability study for flange and mating part(s) to evaluate tolerance.
5. Purposely replicate crack by misaligning components and record force required.

Figure C-3. Cracked Exhaust Manifold Example

Industrial Problem Solving Simplified

The stripped flange bolt problem can help you visualize the forces applied in the assembly of components, as well as any reactive forces that may be present. Figure C-2 shows that there is a potential interference area between the stud shoulder and the mating flange during assembly. This potential interference area is caused when the nut being tightened forces the mating flange into the stud shoulder. The exhaust manifold flange, which was not of proper strength, contained weakened threaded material that was not to specifications. The sketch in Figure C-2 shows where the four threads stripped.

This is further explained in the lower sketches in Figure C-2, which show that the applied force was greater than the resultant resistant forces of the individual threads; these are illustrated by the arrows and force vectors. This imbalance resulted in thread failure. The lower comment box shows examples of the types of observations that can be made in such studies.

The next example, shown in Figure C-3, describes the causes to consider when a component breaks during assembly. If the component material is to specification, you must evaluate other considerations. Hot cracks, impacts, stresses, alignment issues, or excess energy applied are the paramount considerations. In this case, alignment problems were cited. The flange hole was not properly aligned with the mating part, which caused excessive stress. Result: failure.

It should be clear by now that you can identify the causes of cracked or broken components using the methods described in this book. These methods enable you to use a simple concept sheet as a very effective investigative tool.

APPENDIX D

Torque-to-Turn Example

This appendix is included to reinforce several important points that were discussed earlier in this book. These include the need to observe and walk the operating system, the need to observe all possible clues, the need to understand how tiny some clues can be, the need to identify the most prevalent failure location, and the need to ensure that your suppliers agree to get prior approval for any changes they make to materials or processes on their end. Let's look at how all these points play out in a typical manufacturing problem.

Sometimes, two independent conditions can contribute individually to create torque-to-turn problems. In this case, the pistons were supplied by different suppliers.

In this example, an assembly line was used to manufacture two types of engines. The engines had dissimilar faults. One engine had torque-to-turn failures, which means that the force required to turn or spin the crankshaft in the engine assembly was too high or restrictive. This was engine 2 in the on the right side, in Figure D-1, where the torque-to-turn flaw is indicated as "dimensions wrong." The other engine, which was used for comparison, did not have this similar problem. The following discussion explains what we found as we were completing rework to refurbish torque-to-turn, test-engine failures.

Crankshaft Bearing Problems Related To Torque-to-Turn Problems

Pareto chart comparison of crankshaft bearing problems as related to block defects:

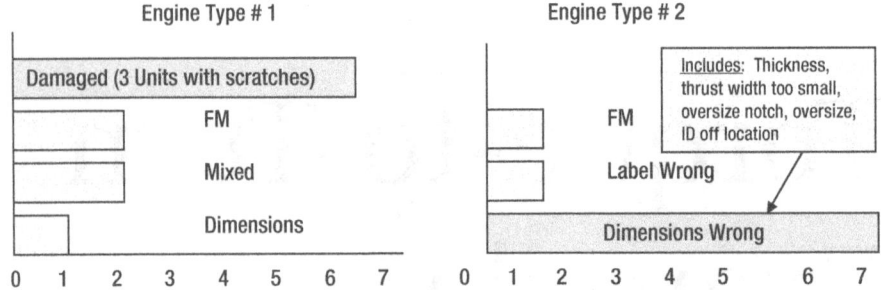

The repair that corrects most of the torque-to-turn failures on the block is the removal and replacement of the ½ round crankshaft to connecting rod bearings. There is no record of any problem issued for damaged or malformed conditions on the Type #2 engine, whereas it is a common fault on the Type #1 engine.

An inspection of the bearing feeder resulted in the removal of a bearing ½ round that had visible scratches. (Operator instructions are to dispose of all damaged units).

Upon further inspection, the bearing appeared to have "shearing burr or tool chatter" on one edge of the bearing at the smaller radius. Samples sent to Metrology showed that they were within specification. However, the high occurrence of dimensional problems indicates that control is lacking and should be improved.

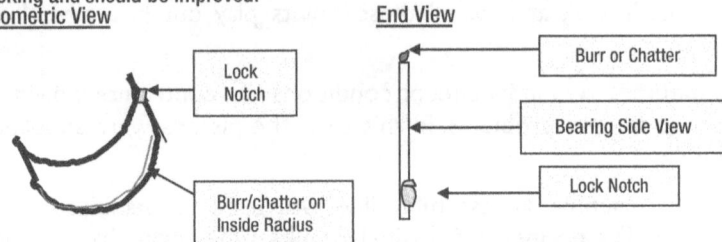

Because of the normal variation inherent in the parts, this condition could be present in some bearings more than others. This may explain why when nothing is found to cause torque-to-turn failures, a change of bearings result in acceptance. Since the bearings are placed by machine, it is believed that the problem is in the part rather than in the assembly. This condition is observed at Rework Station #1 and #2 of Loop #1.

Figure D-1. Engine Torque-to-Turn Failure

The connecting rod bearings were found to be to specification, and there was normal variation inherent in the parts. But there appeared to be a small mark on the lock notch part of the bearing. This mark, observed at a rework station on the assembly line, was present in some bearings more than in others. This may explain why, when nothing was found to cause the torque-to-turn failures, a change of bearings resulted in acceptance. Since the bearings were placed by machine, we believed that the problem was with the part rather than with the assembly.

In the process of the investigation, we also found that there was a predominance of foreign material (FM) present. This was also responsible for creating the torque-to-turn test failures.

Some FM (see Figure D-2) was found between the bearing and the crankshaft pin.

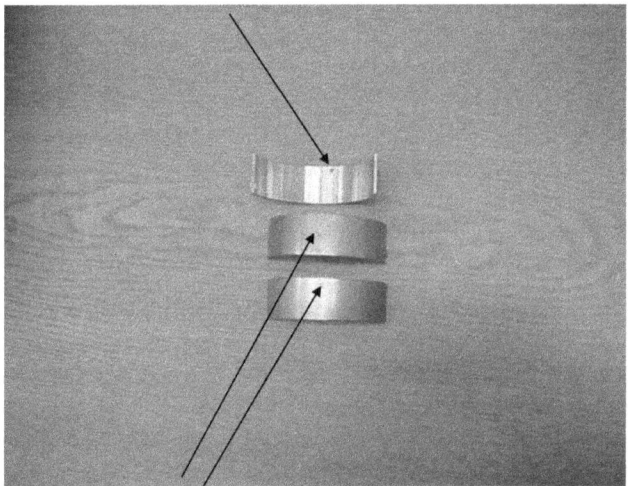

Figure D-2. Photograph of Connecting-Rod Bearing FM

It is extremely important to walk the faulty operation and look for any condition that could contribute to a problem. In this case, we observed that there was a prevalence of chips beneath the piston-assembly station that had accumulated during the manufacturing cycle. The foreign material was later identified as connecting rod flakes that had not been removed from the rods after they had been machined. See results of foreign material in Figure D-3.

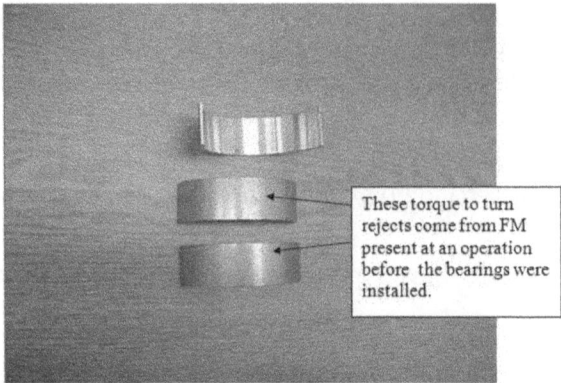

Figure D-3. Photograph of Torque-to-Turn Bearing Rejects

These flakes were due to inadequate DFMEA, PFMEA, testing, and lack of changes to the control plan, as well as the elimination of a cleaning operation by the supplier without permission after the rod design had been changed. The fact that these parts passed the final inspection indicated a failure in the inspection procedures, something that also had to be revised at the supplier.

Figure D-4 shows the old piston connecting-rod assembly on the left and the newly revised piston connecting-rod assembly on the right. The replacement assembly has sloped shoulders, whereas the old one had flat shoulders. This difference in slope allowed the automatic drills in machining to displace by bending a sliver of material on the sloped surface. It had been previously removed when the machining was performed on the flat surface assembly rod. This illustrates the importance that any change, no matter how small, must be reviewed and approved before it can be made.

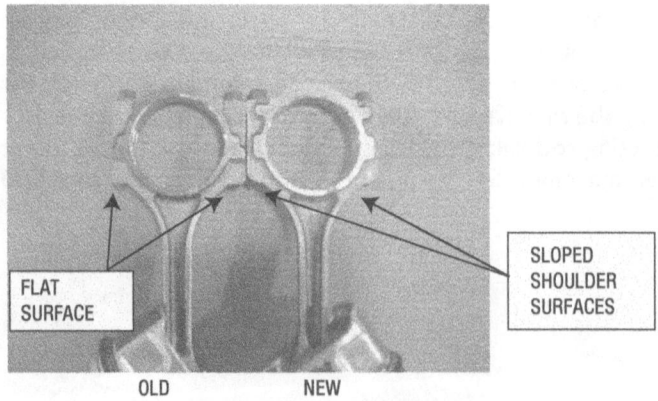

Figure D-4. Photograph of Connecting Rod Modification

During the investigation phase, it was determined that the piston area for the number 5 piston was providing the most problems because of foreign material. The photographs that follow (Figures D-5 and D-6) show the location of the pistons within the assembly-and-transfer fixture after they were unbolted and rotated. This fixture transferred the pistons to an assembler mechanism where they were provided with bearing half rounds before assembly into the engines. This operation was later found to allow some of the flakes to fall into the bearings and the engine bore upon piston assembly.

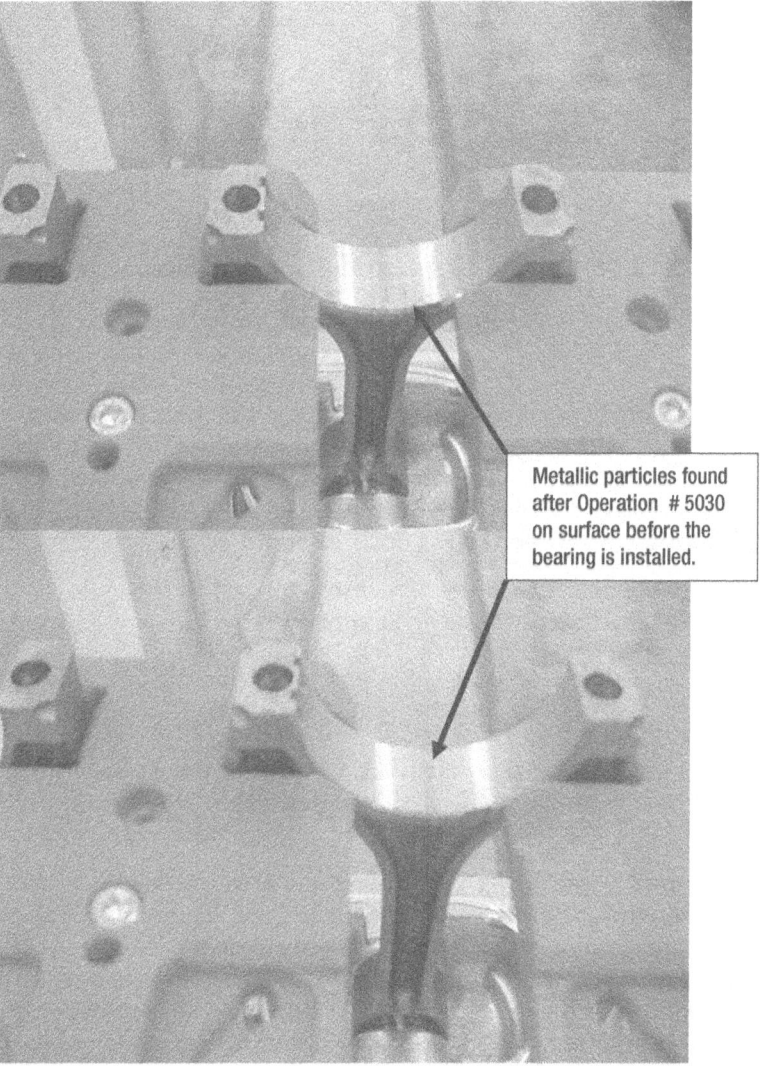

Figure D-5. Piston Assembly Pre-Positioning Fixture

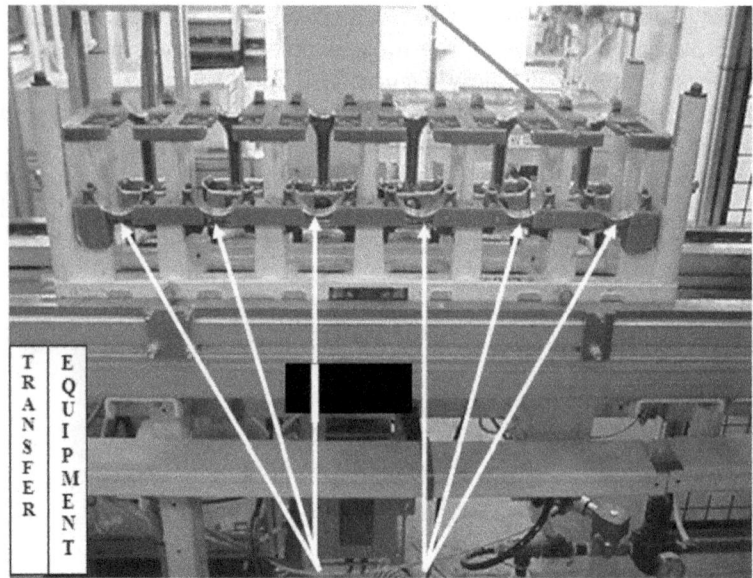

Figure D-6. Photograph of Six Pistons in Assembly and Transfer Fixture

In the evaluation period, we had walked past the piston transfer fixture numerous times without noticing the foreign material (FM) that was under the machine. Only upon walking the floor twice in one day did we notice the FM. During the first walkthrough, the floor had been swept clean; later that same afternoon we noticed the FM under the station.

The pistons were placed into the fixture in sequence as shown, which represented the placement into the engine cavity bore:

```
Bore:          #6      #4      #2      #1      #3      #5
Place Order:   1st     2nd     3rd     4th     5th     Last
```

The biggest foreign material problem was evident from the piston stuffed into bore 5. FM chunks were found at the station before the bearing assembly operation.

There was no FM observed on the half round area on the lower section of the connecting rods before and/or after the bearing was installed into the half round. Even after piston rotation, there was no FM on the bearing in the lower part of the pallet. This was observable up to and including the next sequence operation, which was the insertion of the pistons into the engines. From there, the engines were tested and either passed or were rejected for rework.

Figure D-7 shows foreign material particles removed from the floor under the transfer station.

Industrial Problem Solving Simplified

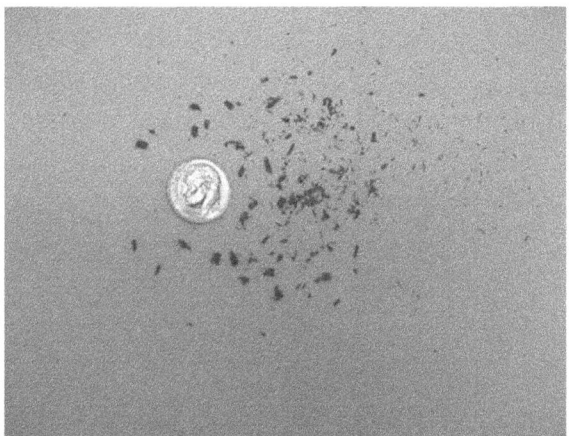

Figure D-7. Photograph of Particulate Found at the Station

We believed this particulate to be powdered metal shavings from the connecting rods. We later proved that this was correct; the newly designed pistons that had a sloped shoulder did not have all the particulate removed after their primary machining operation. The engine plant piston supplier neglected to continue the approved process that had been agreed upon and certified when the contract had been approved. Again, you must scrutinize your suppliers, as they will often attempt to reduce internal costs without prior customer approval.

Many of the methods used to generate clues and to solve problems are apparent in this example. It is not uncommon to find that there may be two or more separate conditions that affect a process and create a flaw. Moreover, whenever materials, processes, or suppliers change, it becomes possible for a negative effect to be produced. Simply walking the problem area, defining the problem, providing photographs, and making visual inspections will generate enough clues to get the investigation started.

APPENDIX E

Sum of Extremes Test

This appendix provides a more detailed and clearer explanation of the sum of extremes test than is contained in Chapter 7. The goal of this appendix is to answer all your questions regarding this type of data comparison. It explains the logic used to design the test, and verifies the means of calculation without too many statistical terms and details. Specifically, this appendix drives toward the acceptance of the comparison of two groups of five as a valid tool that you can use for identifying clues or for verifying that a corrective change is effective.

Figure E-1 shows the ranked values of six samples that come from two different sample groups of three each. The group designated as "better" could be samples produced by an existing manufacturing operation, or they could represent a measured material value that is being used to accomplish that process.

Confirmation of Sum of Extremes Test

There is a means to determine if a significant difference exists between a current and test sample of three of each condition as shown on the next page. Only in condition #1 of the 20 shown below do the three "Better" results fall to the left of the three "Inferior" results. There is clear separation of the 2 different groups that show the Better to the left of the Inferior. Depending on the test requirement, Case #1 could be the better condition if lower test values are required. Case #20 might indicate that a new method or material called "Inferior" has a lower test output than the existing "Better" method. The following example illustrates the above explanation.

If a random sample for testing, such as showing the number of ways to combine 2 groups of 3 items (6 items total) taken 3 at a time, there can be only 20 different ways in which the results of the test can be ranked or described. Condition #1 test results give over 95% confidence that there are significant differences between the "Better" and the "Inferior" parameters.

1.	Better	Better	Better	Inferior	Inferior	Inferior
2.	Better	Better	Inferior	Better	Inferior	Inferior
3.	Better	Better	Inferior	Inferior	Better	Inferior
4.	Better	Better	Inferior	Inferior	Inferior	Better
5.	Better	Inferior	Better	Better	Inferior	Inferior
6.	Better	Inferior	Better	Inferior	Better	Inferior
7.	Better	Inferior	Better	Inferior	Inferior	Better
8.	Better	Inferior	Inferior	Better	Better	Inferior
9.	Better	Inferior	Inferior	Better	Inferior	Better
10.	Better	Inferior	Inferior	Inferior	Better	Better
11.	Inferior	Better	Better	Better	Inferior	Inferior
12.	Inferior	Better	Better	Inferior	Inferior	Better
13.	Inferior	Better	Better	Inferior	Better	Inferior
14.	Inferior	Better	Inferior	Better	Better	Inferior
15.	Inferior	Better	Inferior	Better	Inferior	Better
16.	Inferior	Better	Inferior	Inferior	Better	Better
17.	Inferior	Inferior	Better	Better	Better	Inferior
18.	Inferior	Inferior	Better	Better	Inferior	Better
19.	Inferior	Inferior	Better	Inferior	Better	Better
20.	Inferior	Inferior	Inferior	Better	Better	Better

Figure E-1. Table of Possible Combinations of Two Groups of Three, Taken Six at a Time in Sum of Extremes Test

The evaluator designates the characteristic value that is to be measured and compared to a proposed replacement sample group. The "inferior" sample group represents a proposed replacement operation or a material that is being proposed to replace one that is being used. Each of the six individual samples is measured as determined by the test criteria and ranked with the results in ascending order. If the three better samples have a test score that is lower (in this case that's good; in some tests, a higher value might be more desirable), and not tied with any of the inferior (replacement) sample scores, the data is arranged as shown in Case 1. This arrangement clearly shows that there is a significant difference in the two groups, with the better group showing lower test scores. None of the other ranked results—shown in Cases 2–20—indicate a significant difference in this same ranked relationship between the two groups with any degree of confidence. *Case 1 is the only one that shows that the better group has lower-ranked values than the inferior group.* (Case 20 indicates that the group classified as inferior has in this case a lower score than the better group.)

Only Condition 1 shows that all the better parts rank lower or better than and separate from the inferior rankings. Again, Condition 20 could indicate that the inferior samples are ranked below or more detrimental than the better samples. It all depends on how you set up the relationships. If only one of 20 combinations can produce the results in Condition 1, there is only a one in 20 chance that it can occur when this type of test or comparison is made.

$$1 / 20 = 0.05 \text{ or } 5\%$$

Consequently, if the data arranges itself in this manner after a random sample of the good and bad samples have been tested, you may be inclined to believe that there is 95% confidence that your comparison shows that a significant difference is present. But remember, even when there is 95% confidence that there is a significant difference, there is still a 5% chance that there is no significant difference, and the clue may be erroneous.

100 % – 5 % = 95 % confidence, but 100 % – 95 % = 5 % error potential

For example, consider Figure E-2. It shows the comparison of two samples of fuel pipes. The Cs indicate the brand that was being used and the Ls were a supplier's suggested replacement part, which cost less. The comparison that was made was based on the fact that the proposed pipe had an oval shape instead of a round one. The oval pipe was greater than 0.002 inches on the L part. Consequently, the six individual components were randomized and fed into the assembly line. After assembly, they were tested for leakage. The test showed that there was clear separation of leakage provided by the two different brands. The oval pipes were detrimental. The sum of extremes in this case was six extremes, because there was clear separation of the three Cs and the three Ls, without any ties between the two groups. 3 + 3 = 6.

Appendix E | Sum of Extremes Test

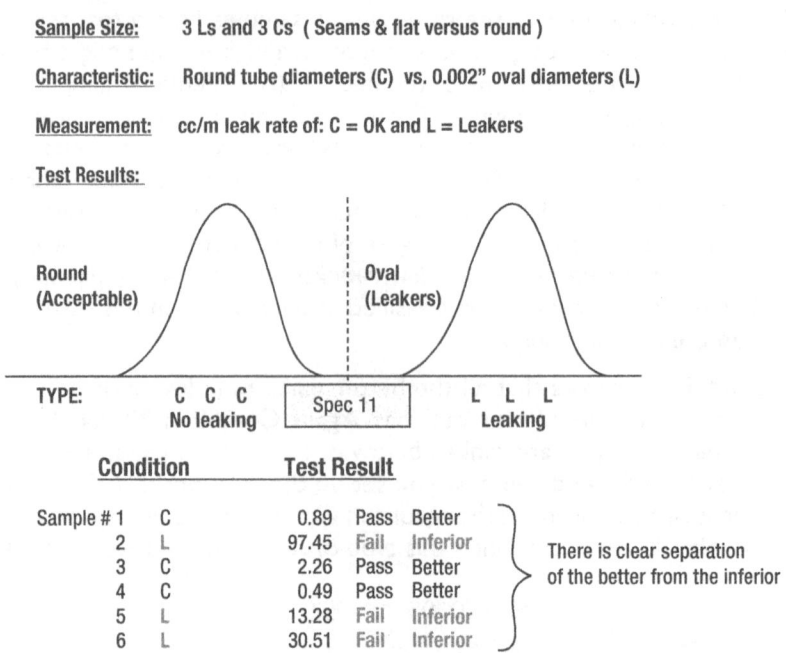

Figure E-2. Sum of Extremes Test Example (Fuel Pipe Leaks); This Is the Equivalent of Line 1 in Figure E-1

So not only can a sum of extremes comparison indicate the difference between two different conditions, samples, products, or results, it can also determine whether a change that is being considered is effective. If there is clear data separation of the groups, you have 95% confidence that the change you made is effective. 95% confidence generally provides good assurance, although you can't establish certainty.

But again, this is only for illustrative purposes. It is better to compare five good and five bad units to generate clues or to validate that the results achieved are truly indicative of the conditions.

This last example used two groups of three, taken six at a time. Now we'll expand that concept to the preferred method of using two groups of five, taken 10 at a time. You can develop this type of table as well. However, in the interest of space, all of the combinations are not shown here. Only ranked values that represent a statistical difference with over 95% confidence are displayed in Figure E-3. These values will still illustrate the rationale for using the recommended five acceptable and five unacceptable samples for measurement, comparison, and ranking.

1. Good Good Good Good Good Bad Bad Bad Bad Bad
2. Good Good Good Good Bad Good Bad Bad Bad Bad
3. Good Good Good Good Bad Bad Good Bad Bad Bad
4. Good Good Good Bad Good Good Bad Bad Bad Bad
5. Bad Bad Bad Good Bad Bad Good Good Good Good
6. Bad Bad Bad Bad Good Good Bad Good Good Good
7. Bad Bad Bad Bad Good Bad Good Good Good Good
8. Bad Bad Bad Bad Bad Good Good Good Good Good

Figure E-3. Possible Combinations of Two Groups of Five, Taken 10 at a Time; They Result in a Sum of Extremes Result of Seven (7) or More

As you can see in Figure E-3, there are 252 ways that five good and five bad units of measurement can be ranked or arranged in ascending or descending order. There are only eight ways in which the ranked data can be arranged so as to create clear separation of the good and bad, with no data overlap or tied values; this method provides a sum of extremes of seven or more.[1]

The sum of extremes in each of these lines is 10, 8, 7, 7, 7, 7, 8, and 10. Note that $8/252 = 0.032$. It is not inconsistent to say that if one of these arrangements is achieved with a comparison of five good and five bad samples (and there are at least seven qualifying extremes), then there is over 95% confidence that there is a significant difference. The different extremes are the addition of the separation of the individuals. See Examples 4 (3 good + 4 bad = 7) and 6 (4 bad + 3 good = 7). Therefore $1 - 0.032 = 0.968$ or greater than 95%.

However, when you're only visually arranging samples, there is a danger that some of the comparisons may overlap, as explained in Chapter 7. Review Figure 7-10, which is in the "Comparison of Individuals in Duos" section. It is therefore imperative that you have a measurement system in place that allows samples to be accurately ranked. Unless all the comparison values in Group 1 exceed all the comparison values in Group 2, there may be excessive overlap in the data and it may not meet the criteria of seven or more extremes. Therefore, all the comparisons of two groups of five taken two at a time must meet the seven or more extremes criteria to be meaningful.

---

[1] If you find this observation to be questionable, I invite you to determine another combination in which the observation is true. All ties between good and bad are not to be included in the sum of extremes and the sum of extremes must be 7 or more.

APPENDIX F

Definitions

Attribute: A quality or characteristic of a part or process.

Attribute Gauge: A device that provides a comparison or measurement that is not based on numerical values. For example: an eye chart.

Calibration: The state in which a measuring instrument is accurate and repeatable for its intended use.

Check: Any assessment of the system methods, tools, or operation that can verify compliance to requirements.

Concept Diagram: A physical sketch illustrating the condition under study.

Consistent Work: The required performance of manpower to tasks using specified materials, tools, methods, operations, and procedures that have been defined to ensure acceptable product or outputs. Components can include and specify safety and environmental requirements as well.

Control Plan: The specified method for organizing and managing a manufacturing or service process.

Corrective Action: The steps taken after a process flaw is recognized to prevent problem recurrence.

DFMEA (Design Failure Modes and Effects Analysis): The tool or document that defines the design considerations that will be applied to a product or process to ensure desired engineering and aesthetic results. The DFMEA is usually the result of a meeting.

Duo: A set of two items, where one is compared to another.

Flaw: A fault having any characteristic that detracts from the intended purpose of a process or service.

Flow Path: The specified route experienced by a component as it travels through a manufacturing or service process.

Appendix F | Definitions

Foreign Material (FM): Any substance that adversely affects a product or process.

Interaction: A phenomenon that occurs when two or more variables react adversely and explosively.

Lockbox: A device that captures defective parts or materials.

PFMEA (Process Failure Modes and Effects Analysis): The tool to define the process considerations that are applied to a product or process to ensure desired engineering and aesthetic results. The PFMEA is usually the result of a quality committee or process group.

Plan of Attack: The steps to be taken to generate clues and to provide problem solutions.

Quality: The fulfillment of customer expectations. If the customer is pleased with the service or product and does not perceive any flaw based on their expectations, it is believed to be of acceptable quality.

Rework: Any operation that is applied to correct a flaw in the product or process.

Sequestered Stock: Product that is captured after a flaw is recognized. It prevents a faulty product or service from reaching the customer.

Sequestering: The method for recalling and holding any product that is in current production, storage, or delivery that keeps the customer from receiving a flawed product.

Sigma: A standard deviation term. This is described as 1/6 of the length of a population distribution. In statistical terms, it is the standard deviation of the mean.

Sporadic Instance: Any spiking occurrence—a rapid and intermittent change in the process flaw production—that does not appear continually as a usual condition.

Sum of Extremes: An evaluation whereby a ranked comparison of as many as 10 samples—two groups of five samples each—provides a clear separation of the data of one group when compared to the other without any tied or overlapping values between members of opposing groups. They are observable at both extremes of the plotted distribution, as explained in the examples in Appendix E.

Tier I and Tier II: The supplier, and the supplier's supplier, respectively.

Validate: To ascertain that tools or results are correct and meet the ascribed requirements.

Variable: A quantity or characteristic that has distinct properties and can be measured or compared because of its inherent nature.

Variable Gauge: A device that provides numerical data. For example: a ruler.

Witness Marks: Any noticeable scratches or discoloration on parts.

Work Procedures: Job procedures that have been specified to perform a service or an operation. This term can be more inclusive and also specify quality, safety, environmental, catastrophe, and communication considerations.

Index

A

Advocates, 160
Analysis tools
  basic analysis
    awareness, 92
    broken conveyor stop mechanism, 92
    corrective action, 93
    hot crack, 93
    problem condition, 90
    problem definition sheet, 91
    problems elimination and prevention, 90
    samples collection, 91
    uses, 90
  bearing caps
    assessment, 105
    castings, 103
    grinder-processing equipment, 103
    insufficient mold hardness, 105
    mold hardness data, 104
    mold wall movement, 104
    trial run, 105
    variables, 103
  clue generation, tests
    assemblies comparison (see Assemblies comparison)
    corrective action, 132
    good vs. bad comparisons, 133
    old material vs. new material, 131
    rating system, 131
    third-cluster bore gear data table, 132
    water filter manufacturer, 131
  comparison of individuals (see Comparison of individuals)
  cracked/broken problem sheet
    dipstick, 152
    dipstick dimple vs. punch design, 153
    qualifier, 151
  forces involved, 101
  fractional analysis
    current samples, 128
    incidental effects, 128
    piston-stuffing concept diagram, 125
    piston-stuffing data sheet, 126
    piston-stuffing interaction table, 127
    unusual random patterns, 124
    variables, 122
  lot identification flowcharts
    assembly process, 142
    consistent work, 140
    control plans, 140
    malformed bearing surfaces, 144
    metal casting, 147
    missing parts, 145
    quality-assurance problems, 140
    scrap items, 141
    unauthorized components, 141
  noise problem matrix, 149
  sound problems, 148
  sum of extremes test
    comparison, 109
    critical difference, 109
    data comparisons, 106
    defectives rate, 108

Index

Analysis tools (cont.)
    explanation and examples, 108
    good values, 107
    groups separation, 109
    leak rate, 106
    proposed replacement values, 107
    ranking, 107
    toy overheating problem (see Toy overheating problem)
    trait peculiarities
        aluminum oxide polishing cloths, 101
        contaminants and foreign materials, 101
        creation steps, 95
        drawing, 101
        individual trait differences, 94, 97
        piece trait differences, 94, 97
        pin supplier, 100
        unusual differences, 94, 97
    visual observation
        confusion, 138
        incorrect valve stems, 137
        meeting call, 140
        piston shoulder differences, 139
        torque failures, 138
Assemblies comparison
    juvenile product manufacturer, 133
    sample separation, 135
    steel tubing, 134
    test gates, 133
    toddler door gate material overlap, 133
Assessment plan development, 76
Attribute data, 67
Attribute, definition, 203
Attribute gauge, definition, 203

B

Balancer assembly noise study matrix, 149
Balance shaft assembly noise, 180
Basic analysis. See Analysis tools
Bent ignition coil connector pins, 29
Bleeding, 41–42

C

Calibration, definition, 203
Check, definition, 203
Clues
    assessment plan development, 76
    collection, guidelines, 80
    condensed problem definition sheet, 81
    conduct assessment checks, 77
    defect diagram
        castings mishandling, 85
        concept sheet, 86
        cracked crankshaft defect location, 84
        fractures, 85
    generation, action items, 73
    manufacture design, 77
    problem definition, 74
    process verification, 75
    react to problem, 77
    unacceptable parts identification, 82
Comparison of individuals
    group data applications, 117
    in duos
        air test, 110
        better vs. inferior, 113
        diameters, 110
        fuel pipe leak test, 115
        leaking fuel rail problem, 110
        length and finish, 111
        maximum pull strength, 112
        ranking indication, 111
        test data separation, 111
        visual signal, 112
    in groups, 115
    regression analysis, 120
Concept diagram, definition, 203
Concept sheet
    Bent Ignition Coil Connector Pins, 29
    crankshaft (see Crankshaft Sensor Operating Zone Analysis)
    definition, 17
    Engine Assembly Routing, 31
    Failure Concept Sheet, 24
    Front Cover Oil Leak Concept Sheet, 20
    Leaks Concept Sheet, 21
    NTF concept sheet, 24

PCB Board Process Map, 33
Poor Fusion Weld Concept Sheet, 18
Condensed problem definition sheet, 81
Consistent work, 156, 203
Control plan, definition, 203
Corrective action, definition, 203
Cracked/broken problem sheet.
See Analysis tools
Cracked exhaust manifold example, 187
Cracked-or-broken problems
check sheet, 183
cracked exhaust manifold example, 187
stripped flange bolt problem, 187
Crankshaft Sensor Operating
Zone Analysis, 27

D

Data collection
attribute data, 67
definition, 63
past and current data, 63
problem sheet uses, 64
variable
calibration, 66
gauge, 66
identifiers, 66
repeat measurements, 67
sensitive measuring instruments, 66
visual evaluation system (see Visual
evaluation system)
Defect diagram. See Clues
Defect scene characteristics, 12–13
Design failure modes and effects
analysis (DFMEA), 48–49, 203
definition, 203
DFMEA. See Design failure modes
and effects analysis (DFMEA)
Downtime, 122–123, 162
Duo, definition, 203

E

Engine assembly routing, 31–33

F, G

Failure Concept Sheet, 24
Fault characteristics
conditional data, 10
defects, 9
flaw, 10
manager role, 10
quantifiable data collection
contrasts of two/more, 13
defect scene characteristics, 12
definition, 12
duos, 14
visual observation, 9
Flaw, definition, 203
Flawed product, 162
Flow path, definition, 203
FM. See Foreign material (FM)
Foreign material (FM), 204
Fractional analysis
operation effect, 170
piston-stuffing fractional (see Piston
stuffing fractional)
Front Cover Oil Leak
Concept Sheet, 20

H

Hot-testing components, 37

I, J, K

Interaction
balance shaft assembly noise, 180
definition, 177, 204
flaming wick, 177
piston-stuffing example, 177
stripped-threads example, 178

L

Labeling, 158–159
Leaks Concept Sheet, 21
Lockbox, definition, 204
Lot identification flowcharts.
See Analysis tools

Index

M
Many-level reviews, 156–160
Materials processing, 162

N, O
Noise problem matrix, 149
No Trouble Found (NTF)
    Concept Sheet, 24

P
Part attribute analysis. See Analysis tools
PCB Board Process Map, 33
PFMEA. See Process failure modes and effects analysis (PFMEA)
Piston stuffing fractional
    operation #2060, 175
    operations 2050 vs. 2060
        correction, 172
        interaction effect, 172
        largest matrix effect, 172
        machining effect, 172
        recommendations, 171
    reject failure, good and
        suspect pistons, 174
Plan of attack
    acceptable and unacceptable
        thrust bearings, 43
    concept sheet, 35
    definition, 36, 204
    evaluation considerations, 37
    part geometry, 36
    prevention
        actions and follow-up, 49
        control plan, 48
        DFMEA, 48
        employee training, 47, 51
        ensuring communications, 47, 59
        ensuring part identification, 47, 57
        ensuring supplier
            compliance, 47, 60
        foreign material, 47, 55
        label problems, 47, 53
        missing operations, 47, 54
        mixed parts, 47, 52
        part coatings preservation, 47, 56
        PFMEA, 48
        procedures, 49
        responsibilities, 47, 49
        spills and sequestrations,
            dealing, 47, 58
        training, 49
    seal conditions
        discoloration, 38
        pattern locations and serials, 38–39
        rework, 38
        seal grease coverage, 38
    tools
        accurate and repeatable, 39
        bleeding, 41
        "double strike", 39
        seal leaking, 40
        "shotgun" approach, 39
        undesirable and desirable
            grease coatings, 40–41
Poor Fusion Weld Concept Sheet, 18–19
Preliminary requirements,
    faults eradication, 167
Problem
    cause, 2
    definition, 2
    gleaning clues
        industrial problems, 6
        problem corrective action
            worksheet, 5
        work sheet information, 6
    inherent in manufacturing
        "cause identification", 3
        customer discontent, 2
        leak test, rejection, 3
        safety problem, 3
    mismatch, 1
    solving conditions, 4
Problem corrective action worksheet, 5
Problem definition sheet, 64
Problem-solving process
    downtime, 162
    fault characteristics (see Fault
        characteristics)
    flaw products, 161
    information, 165
    manufacturing problem, 160
    plan of attack (see Plan of attack)

Index

spreadsheet, 163
temperature, 164
Process failure modes and effects analysis (PFMEA), 204

Q

Quality, definition, 204
Quantifiable data. See Fault characteristics

R

Rework, definition, 204

S

Sequestered stock, definition, 204
Sequestering, definition, 204
Sigma, definition, 204
Solve problems, 167
Sporadic Instance, definition, 204
Spreadsheet, 163
Stripped flange bolt problem, 187
Sum of extremes definition, 204
Sum of extremes test
    better vs. inferior test, 200
    data arrangement, 199
    ranked data arrangement, 201
    ranked values, samples, 197–198
Supplier, 162

T, U

Tier I and Tier II definition, 204
Torque-to-turn problems
    bearing rejects, 192
    connecting rod bearings, 190

connecting-rod modification, 192
engine plant piston supplier, 195
foreign material particles, floor, 194
piston assembly pre-positioning fixture, 193
pistons in assembly and transfer fixture, 194
torque-to-turn flaw, 189
Toy overheating problem
    battery compartment redesign, 98
    battery temperature pattern, 98
    consistent work uses, 100
    finance managers, 99
    iron oxide supplier, 100
    supplier role, 99
    unauthorized procedural changes, 100
    units testing, 98

V

Validate, definition, 204
Variable, definition, 204
Variable gauge, definition, 205
Visual evaluation form, 68
Visual evaluation system
    construction, 69
    grease, visual acceptance system, 71
    porosity, visual rating, 69–70
    standard ratings, 69
    visual evaluation form, 68
    visual rating sheet, 69

W, X, Y, Z

Witness marks, definition, 205
Work procedures, definition, 205

Get the eBook for only $10!

Now you can take the weightless companion with you anywhere, anytime. Your purchase of this book entitles you to 3 electronic versions for only $10.

This Apress title will prove so indispensible that you'll want to carry it with you everywhere, which is why we are offering the eBook in 3 formats for only $10 if you have already purchased the print book.

Convenient and fully searchable, the PDF version enables you to easily find and copy code—or perform examples by quickly toggling between instructions and applications. The MOBI format is ideal for your Kindle, while the ePUB can be utilized on a variety of mobile devices.

Go to www.apress.com/promo/tendollars to purchase your companion eBook.

All Apress eBooks are subject to copyright. All rights are reserved by the Publisher, whether the whole or part of the material is concerned, specifically the rights of translation, reprinting, reuse of illustrations, recitation, broadcasting, reproduction on microfilms or in any other physical way, and transmission or information storage and retrieval, electronic adaptation, computer software, or by similar or dissimilar methodology now known or hereafter developed. Exempted from this legal reservation are brief excerpts in connection with reviews or scholarly analysis or material supplied specifically for the purpose of being entered and executed on a computer system, for exclusive use by the purchaser of the work. Duplication of this publication or parts thereof is permitted only under the provisions of the Copyright Law of the Publisher's location, in its current version, and permission for use must always be obtained from Springer. Permissions for use may be obtained through RightsLink at the Copyright Clearance Center. Violations are liable to prosecution under the respective Copyright Law.

Other Apress Business Titles You Will Find Useful

The CPO
Schuh/Strohmer/Easton/
Scharlach/Scharbert
978-1-4302-4962-7

Disaster Recovery, Crisis Response, and Business Continuity
Watters
978-1-4302-6406-4

Exporting
Delaney
978-1-4302-5791-2

The Handbook of Financial Modeling
Avon
978-1-4302-6205-3

Improving Profit
Cleland
978-1-4302-6307-4

Makers at Work
Osborn
978-1-4302-5992-3

Inventors at Work
Stern
978-1-4302-4506-3

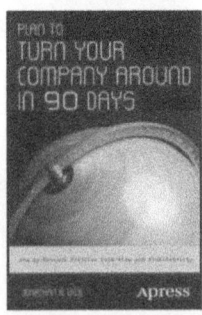

Plan to Turn Your Company Around in 90 Days
Lack
978-1-4302-4668-8

How to Get Government Contracts
Smotrova-Taylor
978-1-4302-4497-4

Available at www.apress.com

GPSR Compliance

The European Union's (EU) General Product Safety Regulation (GPSR) is a set of rules that requires consumer products to be safe and our obligations to ensure this.

If you have any concerns about our products, you can contact us on

ProductSafety@springernature.com

In case Publisher is established outside the EU, the EU authorized representative is:

Springer Nature Customer Service Center GmbH
Europaplatz 3
69115 Heidelberg, Germany

www.ingramcontent.com/pod-product-compliance
Lightning Source LLC
LaVergne TN
LVHW040736250326
834688LV00031B/323